Night Vision

Night Vision

In search of the true dark

JEAN SPRACKLAND

JONATHAN CAPE
LONDON

1 3 5 7 9 10 8 6 4 2

Jonathan Cape, an imprint of Vintage, is part of the
Penguin Random House group of companies

Vintage, Penguin Random House UK, One Embassy Gardens,
8 Viaduct Gardens, London SW11 7BW

penguin.co.uk/vintage
global.penguinrandomhouse.com

First published by Jonathan Cape in 2025

Typeset in 11.1/15.2pt Adobe Caslon Pro by Six Red Marbles UK, Thetford, Norfolk
Printed and bound in Great Britain by Clays Ltd, Elcograf S.p.A.

The authorised representative in the EEA is Penguin Random House Ireland,
Morrison Chambers, 32 Nassau Street, Dublin D02 YH68

A CIP catalogue record for this book is available from the British Library

ISBN 9781787334236

Penguin Random House is committed to a sustainable future
for our business, our readers and our planet. This book is made
from Forest Stewardship Council® certified paper.

For my children,
made of darkness and light

Contents

Preface: Into the Dark

Darkness is hard to see. One starless night last winter I stood alone in the lane and thought about the spell it casts on a place, making it feel different not only to my eyes but to all my senses. I knew my surroundings well but I couldn't see them at all; I couldn't even see where I ended and they began. The air felt thicker than usual, and sound travelled mysteriously: the scratching of dry leaves could have been a shrew in the ditch beside me or a roe deer across the valley. The more I considered it, the more surprising it seemed that something so simple could have such a transformative effect. Come to think of it, I asked myself, what *is* darkness?

Surely the question was absurd. Darkness is the absence of light – the dictionary tells me this, and the nights in this place attest to the truth of it. But how can a mere absence have such a powerful impact on my body as I balance, orientate myself and move through space? How can it alter the way my mind works as I pick up atmosphere and perceive reality?

Darkness is hard to see for other reasons too. A cloak of metaphor is thrown over it, disguising its true identity. Even as I experience it firsthand – stumbling from bed in the small hours, or going down the steps into a cellar, fumbling for the light switch – it can slip into invisibility, leaving behind an image, a symbol. The pleasures of the dark – a clear night sky, a gathering of friends around a bonfire, a space for creative thought, intimacy or rest – leave hardly a trace in the language. Most of the time it's taken as a way of articulating a range of abstract states and conditions, all of them negative: death, oblivion, ignorance, depression, satanism and the occult, war, sin and so on. Darkness is not any of these things, but it becomes a proxy for them.

If you type the word into a search engine, you will trawl up an enormous seething net of results like these. It goes on for hundreds, thousands of pages. The gloom is pierced by occasional shafts of light. Mind-blowing articles about dark matter, dark energy and the dark side of the moon (none of which is actually dark). The fact that an Old English word for darkness, *heolstor*, also meant 'hiding place' and eventually became 'holster'; and that the darkest material in the world is called Vantablack, which in the form of sprayable paint is exclusively licensed to the artist Anish Kapoor. But the rest is desolation, real and imaginary. As you scroll and scroll, you might think of something the poet Kathleen Jamie wrote: 'Pity the dark: we're so concerned to overcome and banish it, it's crammed full of all that's devilish, like some grim cupboard under the stair.'

There's no ignoring the bad in the world, no way of avoiding it. We live in a time of manifold crisis, and each unfolding threat, loss and suffering is reflected and refracted down an endless digital hall of mirrors, twenty-four hours a day. Sometimes despair seems the only possible response. But it's easy to forget that none of it is the fault of the dark. As poets know well, metaphor exerts a mighty influence on thought and feeling; the duality of 'light good, darkness bad' is so deeply embedded in the collective consciousness it has come to seem like an essential truth, and plays a part in shaping our attitudes towards the real thing. No wonder we keep turning on more lights.

The truth is that darkness is a necessary condition for life on earth, and at this time it is imperilled and in need of protection. The nocturnal world teems with organisms that hunt, feed and reproduce only in the dark. And when I walk through city streets under a glowing sky, I'm conscious that I need it too. Yes, I love light and colour, long summer evenings and dazzling winter sunshine, but darkness offers gifts of its own. How can we begin to prise the real thing free of its negative associations, which are as old as human thought itself?

A few years ago, we bought an isolated cottage in the Blackdown Hills, close to the border between Somerset and Devon. In this small valley, life flows at a slower pace and the nights can be long and vertiginously deep. We have no internet connection or mobile signal, nothing to mediate between me and my surroundings. I spend a lot of time outdoors, where I watch the sky, noticing the shift and interplay of light and dark, how different everything looks and feels after dusk. These dark nights are an unanticipated gift of the place. I want to immerse myself, to delight in them, to feel at home in them. But I often find myself holding back. Some part of me remains alert and vigilant, scanning my surroundings, ears pricked for unfamiliar sounds. I love the dark, but it can be so deep I'm afraid of toppling in and never being able to climb out again.

There is darkness in urban environments too, but no risk of toppling in. Cities are heady and glittering places – it's part of their attraction. But you can end up feeling trapped and hemmed in by light, glaring from streetlamps and buildings, shop windows and billboards, spilling upwards and drowning the stars. It's almost possible to forget that anything exists beyond the smoky orange glow of the sky; the city comes to feel like the universe. It's a twenty-first century reversion to the geocentric model, with us at the centre of everything and darkness nothing more than our own shadow.

If darkness has me under a double spell, I'm not the only one. We humans have a complex and contradictory relationship with darkness. We have always feared it, and made great efforts to mitigate it, even to blot it out, bringing our technologies and ingenuity to bear on the challenge. At the same time – especially for the city dweller who remembers the less polluted skies of their childhood – the truly dark sky can represent a kind of lost innocence, closeness to nature and a state of belonging in time and space. Travelling to some remote spot offers an opportunity to regain that state, at least temporarily. To stargaze, and to observe the rich nocturnal life that prevails beyond the range of streetlamps. To sit at the mouth of a tent and listen to the sounds of the night – owls calling, branches in the wind, unknown creatures barking and rustling – without access to the usual visual cues. To feel the landscape re-enchanted, and the self with it. Dark sky tourism is booming. But not everyone can get away to such places, and for some people the sight of an unspoilt sky is a distant memory, while others have never experienced it at all. Darkness, once a birthright, is becoming a commodity.

On the train from London or Manchester, I imagine a conversation between city and country. *Nothing stays the same for long*, says the first. *Some things are eternal*, says the second. The dark is one of the eternal things. It alters my sense of space in a contradictory way, confronting me with incalculable distance while bringing me into closer relationship with my environment and the life that surrounds me. As I come down the lane I realise with a jolt of the heart: *I'm here!* The feeling of awe cuts straight through to somewhere cramped and depleted inside me, reopening a space for awareness of self. It reminds me that I am not a bundle of functions, anxieties and to-do lists, not a device plugged in to a giant computer, but a body in space: entire, flawed, mortal, one among an infinite variety of living things, all of us here together – afloat or shipwrecked – in this one miraculous moment.

All this comes from an absence of light. Standing there that still January night, I thought not only about darkness but also about silence, which of course the dictionary defines as the absence of sound. It occurred to me that they are similar, these two great absences, both of which feel like actual states, like huge rooms furnished with sensation. I thought of John Cage's famous composition *4'33"*, in which the orchestra assembles but never plays a note. Every time it's performed, the piece demonstrates that true silence does not exist. The 'music' the audience hear as they sit together in the concert hall is incidental: a distant siren, a cough, a floorboard creaking and settling, a gurgle of digestion, the in and out of their own breathing. It struck me that the same logic applies to darkness: that what we call *absence* is really just *less*. After all, the lane was getting a little less dark as I stood there and my eyes gradually adjusted. The orchestra of moon and torch had not played a note, but now I could see my own gloved hand, the hedge, the footpath sign. What's more, it's relative; my night vision is very limited compared with that of an owl or a tiger, and for the giant ostracod, which has parabolic mirrors for eyes, even the impenetrable dark of the deep ocean is not total.

Darkness is a human construct, subjective and partial. It's possible to love it and fear it at the same time. It complicates our knowledge of the world, introducing ambiguity, softening hard edges with uncertainty. We think we know it well, and much of the time it goes unnoticed: too commonplace, too obvious to be worth considering. What can there possibly be to say about it? I tried to puzzle it out, while the clouds slid over the stars and my own white breath drifted and dispersed. A cloak of metaphor concealing an absence that isn't really an absence – nothing could be more slippery and elusive.

On that winter night I decided to go in search of the true dark, in which we all once belonged, before we were dispossessed. So this is a book about actual rather than figurative darkness: how it affects us and why it matters; about some of the ways in which it has been explored and interpreted; and what it has meant to me in different places and different states, in childhood and adolescence, in childbirth and illness, love and grief. About the role it plays in adventure: not territorial conquest and the planting of flags, but the unpredictable, messy exploits of people who set off to explore the dark of the high Arctic or the deep cave, and discover truths about themselves in the process. And my own modest versions, much smaller in scale and closer to home: adventures anyone could have if they wanted, because darkness can make the most ordinary activity feel adventurous, touching familiar places and things with strangeness, answering each question with another question.

These small but necessary adventures begin with opening a door and stepping through it. You can't see a thing: you could be anywhere. The door might open outwards onto a city street or a country lane; or inwards on a house, a hut, a dream. Any place will do. Step over the threshold, into the dark, and feel your way.

Darkling

Cottage

At seven on the morning of the winter solstice, it's still night. By seven-thirty, there is the faintest pallor on the horizon where the field rises to the east. The trees are restless, as if sensing the balance is tipping on the fulcrum. The weeping beech in front of the house stirs as if the darkness itself is moving it. A wavering light travels between the trees at the end of the orchard: a torch being carried along the garden of our neighbours' cottage at the bottom of the lane, no doubt by one of the girls collecting eggs for breakfast.

I can't imagine being afraid of the dark in the mornings. At the onset of a winter night it stretches and lengthens ahead, as if daylight has been cancelled out. It can raise the old fear of cosmic abandonment, set out so graphically by Byron in his poem 'Darkness':

> *The bright sun was extinguish'd, and the stars*
> *Did wander darkling in the eternal space,*
> *Rayless, and pathless, and the icy earth*
> *Swung blind and blackening in the moonless air*

But now, with that barely perceptible glimmer in the east, such apocalyptic thoughts would seem absurd. And getting up on these dark mornings offers a simple gift: the opportunity to watch it get light. The transition is more than merely beautiful; it's a profound and affirmative thing to witness.

3

I have always loved winter: the shortness of the days, the sense of drawing inwards, of gathering. As a child I needed that season of retreat and interiority, and the permission it granted to stay indoors and read. In my adult life I have noticed that my sense of creative openness or porousness is at its most intense in October and November, as the nights lengthen. But I know that for some people, it's a depressing and difficult time which they endure rather than enjoy. Who knows why we all perceive it so differently. One friend welcomes it as a reminder of her own fearlessness, learnt in childhood when she went alone across a farmyard at night to tend to the animals. But another tells me that the dark of winter robs the world of colour, and that seeing colour is crucial to her happiness and wellbeing, and a third says vehemently that he hates the dark because to him it means just one thing: death.

Many dark metaphors have a sense of finality about them. When Vladimir Nabokov wrote that life is 'a brief crack of light between two eternities of darkness', he was drawing on two of the great symbols at once: darkness as oblivion, and darkness as ignorance. I cannot remember the time before I was born, nor can I know what it will be like to be dead. Earthly consciousness is richly multisensory, but the extinction of that consciousness is almost always imagined in terms of vision. The dread is of endless darkness from which there can be no escape.

In life, however, dark is always followed by light, and light by dark. The reality of my lived experience is change, not stasis. Regular patterns – day and night, the seasons, migrations and cycles of life – are what anchor me to this world; fear of their breakdown is voiced everywhere, and not only in our own time.

But look – a thrush – or is it a blackbird? My first bird of the day, dimly visible against a background of grass that is now faintly green.

In this place, the passage of time can feel slippery and equivocal, like the ground beneath our feet. We live on the springline, the point at which permeable greensand and impermeable clay subsoils meet, squeezing water out sideways, where it wells up as springs, and forms mire and quaking bog. The land here is too wet to grow crops. A few cattle, or a few sheep, that's about it. Some vegetables, an apple tree or two. These basic conditions remain unaltered to this day. Houses are few, and it's not easy to sell them.

The clock has ticked on, however, and life in the valley has changed, not least at night. Previous occupants of this place knew its darkness far more intimately than I do. Electricity was not installed until the 1960s; before that, the cottage was fitted with oil and then paraffin lamps, but throughout most of its history it was lit by 'rats' tails', homemade lamps made from rushes dipped in tallow. The children would be sent out to gather the rushes, which grow abundantly in the wet and boggy ground, and they were soaked and stripped and spread out to dry on the hearth before being cut into lengths ready for use. The tallow was a by-product of the raising of sheep and cattle for food: animal fat, carefully saved and eked out, pressed into little dishes made of clay or stone. Those early occupants made do with very little, and that simple wick in a saucer of fat was not very different from the kind of lamp in use three thousand years earlier. Just as well they had easy access to the raw materials, since the cost of readymade candles would certainly have been beyond their means.

For most of human history, lighting depended on fuel

derived from edible sources – fish oil, whale oil, tallow, vegetable oils such as coconut, olive and castor – and there was a tension between the need to eat and the need for illumination. Before industrialisation, the local ecology determined the availability of fuel, and the people who had the readiest access to light lived not in cities but in places where there was an abundance of cattle, sheep or oily fish.

The people living in this valley were very poor, but some animal fat could usually be spared for the making of lamps. In times of shortage, however, food came first and homes were left in the dark. The introduction of inedible fuels was revolutionary. The breakthrough came in the nineteenth century: first oil of turpentine, which had the disadvantage of being highly explosive, and then paraffin, a more stable alternative. Crucially, it was not edible. The competition between lighting and food began to subside.

The first time we saw this place, we came by train and took a taxi from the nearest town. The driver, navigating the steep, narrow, twisting lane into the valley, was appalled. 'No one lives here,' he told us, firmly and repeatedly. It was a bleak February day, and the estate agent, impractically dressed, bouncing on his heels to keep warm, could hardly believe his luck; the place had been on and off the market for years, the owners long gone, and it was looking like a hopeless case.

We pushed the gate and it dropped from its hinges. We stood and stared. The cottage, long ago rendered and roofed with tiles, was ensnared in a thicket of brambles as stout as rusted railings. Sleet ran straight off the roof and pooled on the broken concrete. Electricity wires trailed on the ground between the house and some kind of outbuilding. 'No,' I said, but I spoke with the desperation of a lover already caught in the trap. *No.* In the lane behind me, the taxi driver was still attempting to turn round, clutch burning, wheels spinning uselessly in the mud.

We liked the feeling of remoteness, the hills, the quiet. We knew we had a lot to learn about living in such a wild and wet bit of country. But we didn't think at all about the nights until a few months later, when the deal was signed and sealed and we came with sleeping bags and camped out in the empty house. We had no curtains yet, and when I woke in the night the square of the window frame held so many stars I thought I must be dreaming.

Sylvia Townsend Warner, who devotes no more than a page and a half of her book *Somerset* to the Blackdown Hills, nevertheless captures them well. 'Lanes in every degree of greenness and forsaken windingness branch off the roads,' she writes, 'lanes which are rutted with cart tracks, or closed off by sagging gates, lanes whose thick hedges grow out of stony banks, lanes that disappear downhill under green tunnels. If you follow them they will entangle you in further lanes, or take you, with an artful circumvention of contours, to some farmhouse perched halfway down a hillside, with a climbing field behind it dotted with sheep.' She was writing in 1949, but her description is as accurate now as then. The 'forsaken windingness' has so far proved an insurmountable obstacle to the provision of broadband; a few years ago there was talk of it, and a contractor came and painted cryptic symbols at intervals on the roads, but shortly afterwards it was pronounced impossible. A few of the symbols survive, faded to the point of archaeology.

To begin with, this cottage was two tiny dwellings, built on a field to house farm labourers and their families. The men themselves built their homes, using whatever they could find in the fields they tended: chert and rubble for the walls, straw to thatch the roof. It was heavy work. An agricultural gazette remarked of local conditions at the time: 'Little can be said that is favourable. The land is in the hands of men of no capital, who employ hardly any labour.' Scraping a living on the springline was always tough, but the wider economics of agriculture varied over the years and perhaps it was during an upturn, on a spike of optimism, that the decision was made to build.

Each of the two dwellings was dominated by a stone ingle-nook in which a fire was kept burning. That domestic fire was vital for warmth, cooking and illumination, and at night it was all that stood between those families and the blinding darkness outside their doors. It was kindled with dry bracken and fuelled by wood, gathered where it fell from the trees or harvested through pruning and coppicing. If there was a shortage of firewood, it was supplemented with furze or gorse. The gorse must have been awkward and scratchy to harvest, and once it was gathered into faggots or 'blackjacks' it had to be left for six months before it was ready to burn. It still grows profusely in the nearby fields, putting forth its chromium yellow flowers and warm coconut smell, but no one harvests it now. Bracken, too, flourishes, dies back and goes to waste, though it was for centuries considered an important crop, providing bedding for animals as well as kindling for the fire.

I don't suppose Sylvia Townsend Warner hung around here after dusk. When she noticed the dark creeping up one of those green tunnels towards her, she'll have jumped in her MG, back to the lights of town and a gin in the hotel bar. Who could blame her? In any case, there's plenty of darkness to be found in this landscape, even by day. It's held in the steepness and narrowness of the valleys, the sunken lanes and thick hedges, the black mire of the springline and the old turbaries where people used to dig peat for their fires. It settles in the ancient woods, in the dense shade of oak and beech, until a storm uproots a few trees and sunshine pools on the ground where a space has opened up in the canopy. Wherever a patch of the woodland floor is exposed to the sky, there is a sudden bloom as long-dormant plants take advantage of the light. But elsewhere the woods keep their shadows, along with the ruins of abandoned settlements – Quants, Hawks Moor, Jacob's City – unknown to the intrepid walker until they trip on a piece of masonry in the undergrowth. The landscape has changed slowly here, and the past is always there at the edge of the vision.

The dark nights feel to me like a privilege, but I'm sure they had a different character for the families who congregated here in the winter. Census returns show that at times there were eight or nine people living in each of the adjoining cottages, and the downstairs room with its inglenook would have been very crowded. Local records tell of births and deaths within these walls; sex out of wedlock and with the wrong brother, a ghastly accident, a man in uniform walking away up the lane to war. Sometimes on windy nights I think I hear voices in the chimney, but I can't tell what they're saying. I unlatch the little iron door and put my hand into the old bread oven, feeling around in the dust for something left behind. There are strange marks scratched into the stone over the inglenook, but who knows what to read into them?

On midwinter nights, I think of the young daughters of the family. Girls hereabouts contributed to the family subsistence by doing piecework sewing gloves. They began to learn the trade at the age of seven; it was very fine work that required young eyes, and much of it was done by firelight and rushlight. A local clergyman warned: 'It makes them bad wives, for it is a kind of sewing useless for needlework or repairing clothes.' But where money was scarce every pair of gloves mattered. Two dozen pairs a week would bring in four shillings, making an important contribution to overall income.

They, too, are right at the edge of the vision, those long-gone girls. If I keep very still, they might come into focus, sitting bent over their work, making gloves they can never hope to wear themselves. They chatter and giggle as they poke rows of holes in the lambskin with a little awl, and make their neat stitches with a special three-cornered needle, all by the sooty light of a rat's tail.

Library

Many years ago, I went on a long retreat in a big old house in Scotland, a whole month with nothing to do but write. I had a contract for a book of poems, and I needed to make progress. My children had come to the station to see me off. They roller-skated up and down the platform, and I leant out of the train window and watched them.

The taxi pulled up on the sweep. It was late November, a lashing wet night. I rubbed the car window with my sleeve and saw a door open in the wall and yellow light spill out onto the gravel.

The house creaked and smelled of polish and roasting meat. Its halls and passages were glowing and shadowy. We trod them softly, not to disturb each other's work.

The glen was narrow and deep, and dusk came at three in the afternoon. I would put on my boots and walk up and down, up and down. The water carried grey pages that had escaped from the paper mill upriver. They would catch on low branches and hang there, still harried by the rushing water. In my dreams they were poems torn off and swept along, speared on branches, always just out of reach. I did try scrambling down the bank to rescue one, but it was so steep and slippery I was afraid for my life.

Indoors, I sat idle at a black window until I couldn't bear my own reflection any more. My room was panelled with oak and a fire smoked in the grate. There was a huge four-poster bed with wine-red drapes. All day I sat at the table under the window and nothing happened. I had sharp pencils, the kind I liked, and a new notebook. I got up and paced the room. Nothing. Back to the window. Nothing. I put my head on the table and wept with fury. I did not want to be there, I could not write any poems, I hated poems, I could not do it.

When it became impossible to sit there any longer, I took a torch and crossed a courtyard to the tiny library. It was icy cold inside. I fumbled for the switch. The room was crepuscular, mostly populated by shadow and my own white breath. I reached at random for a book. I told myself I didn't care what it was at all. It was economic theory. I put it back. Now I pushed the little ladder on its wheels and climbed it, reaching into the gloom and feeling with my fingertips for something to save me. I touched the spines of the books, and the shelf gritty with dust that was not really dust but disintegrating paper and glue and bits of insect wing. I prised out a book and climbed down with it. I flicked the switch and left the library to its winter dreaming.

I was right the first time, any book would have done – even the economic theory. All I needed was to go into the dark and bring something back. But what I had in my hands now was an illustrated book of saints. In my smoky room I turned a page and there she was, unmistakeable – the woman I had half seen, half imagined the day before, at the river's edge, squeezing water from an infant's garment and shaking it out. I had seemed to catch her eye, and she had stepped back barefoot over the mud and into the mouth of the cave and was gone. I reached for my notebook.

I still have in my possession a copy of T. S. Eliot's *Four Quartets* with the school library label posted inside the front cover. There's a list of pupils' names and the dates on which it was issued. I have written my name in my best handwriting, and looking at it now I remember the ink pen I had at that time, a birthday present from my grandparents. This might have been the first time I'd used it. My writing is careful, self-consciously neat but with a couple of little flourishes – the wavy line over the J, the extravagant loop of the y in my surname. It's the start of a new school year, and this is the style I've decided to adopt. I remember the bottle of blue Quink I used to carry around in my school bag, the ritual of filling the pen by prising up the silver lever on the side and letting it drink it in.

As a librarian's daughter I feel a bit ashamed about my failure to return this book (and several others). It occurs to me that I still could, though the school has a different name now and times have changed, and it seems unlikely that they still study *Four Quartets*. In any case, I could never go back to that place. And I can't part with it now, mainly because it's so entertaining to read my annotations, in which I earnestly point out the literary devices and underlying meanings of the text. The most amusing – the one I'm looking at now – is on page 27, next to the opening lines of part three of 'East Coker':

> *O dark dark dark. They all go into the dark,*
> *The vacant interstellar spaces, the vacant into the vacant,*
> *The captains, merchant bankers, eminent men of letters,*
> *The generous patrons of art, the statesmen and the rulers,*
> *Distinguished civil servants, chairmen of many committees,*
> *Industrial lords and petty contractors, all go into the dark*

In the margin I have written, not in my blue Quink but with a well-sharpened 2D Staedtler pencil, the single word 'Death'.

Just as an apple stood for temptation, and autumn for the passing of time, so I learnt that darkness in a poem meant Death with a capital D. It wasn't a bad rule of thumb. Sometimes it was given a radical spin: we read Henry Vaughan's poem 'The Night', for instance, which envisages a God made of 'deep but dazzling darkness' and an eternity spent in that paradoxical state. This seemed a useful addition to the vocabulary of dark: just as light can be 'blinding', I think darkness – the actual kind – can indeed also be 'dazzling'; both adjectives carry a similar sense of being confounded and overpowered. (More mundanely, I find that Dazzling Darkness is also the name of a Dulux paint colour, but on the website it looks unlikely to dazzle – it's a brownish grey, reminiscent of school skirts.)

Darkness has been a recurring presence in my own poems. The disorientating dark of the sand dunes at night, pulsing with the sound of natterjack toads; the tumultuous dark of a bedroom with a barn owl trapped in it; the darkness of outer space, of a garden shed, of a city doorway. I don't entirely know why it crops up so often, or what I intend it to signify. The subliminal, perhaps. The ulterior.

I certainly don't want to call on it to play the role of bad actor. 'When any poet would describe a horrible tragical accident,' wrote Thomas Nashe in 1594, 'to add the more probability & credence unto it, he dismally beginneth to tell how it was dark night when it was done, and cheerful daylight had quite abandoned the firmament.' A mischievous remark, delivering quite a sting even now. It is dangerously easy to fall into cliché where light and dark are concerned.

My thinking about poetry has circled back again and again to the immense power of darkness and its capacity to speak of loss, vacancy and oblivion. When I embarked on this book, I anticipated writing about some of the great poems of darkness. I compiled a list, and constructed a tottering pile of books that bristled with Post-it notes. But it turned out that most of these poems were not particularly concerned with darkness itself; they usually spoke through it rather than of it. What I was looking for as I read and re-read was actual darkness, made present and palpable, full of potential, erotic, numinous or revelatory.

Where I found it, it was mostly fleeting and fragmentary. Theodore Roethke, sketching 'A night flowing with birds, a ragged moon'. H. D. observing how, in the shadows of dusk, 'each leaf / cuts another leaf on the grass'. Tomas Tranströmer, in his poem 'The Couple': 'They switch off the light and its white shade / glimmers for a moment before dissolving / like a tablet in a glass of darkness'.

On the other hand, the word itself is so highly charged with metaphoric power that it does a lot of work on its own, especially if we hear it again and again. In 'Bavarian Gentians' by D. H. Lawrence, the word 'dark' repeats and repeats, as if he is dipping a brush into paint – perhaps a tin of Vantablack he's nicked from Anish Kapoor – and applying it layer upon layer. With every repetition, the darkness thickens and intensifies. The poem becomes an incantation, drawing us 'down the darker and darker stairs, where blue is darkened on blueness' – not into the subterranean gloom of the coalmines he remembered from childhood, but the deeper underworld he knew was close at hand, because the poem was written just months before he died of tuberculosis.

Darkness is a condition in which poetry can begin. It's the space in which the poet moves around their subject, in which they reach out and touch what they cannot see. An image called forth by memory, imagination or the two working so closely together you can't feel the join. A word, or a little chain of words, loosely connected, as if caught in a cobweb – you could accidentally brush against the web and the words would scatter: insect, leaf skeleton, bit of blown polystyrene.

How easily we drift into the inky waters of metaphor. Poor old darkness is again reduced to a figure of speech, this time a cipher for that ethereal condition in which a poem begins. I try to say it another way, but I keep coming back to this, because it's the nearest I can get to answering that eternal question: *Where do poems come from?*

Concentrating hard, trying to visualise something, we instinctively close our eyes. It helps to block out the distraction of the here-and-now, which is so insistent in its demand on our attention. Vision is the busiest of the five senses, and to shut it off for a moment can help us focus on the thing we want to call to mind. So it is with darkness and poetry. If the poet holds their subject under a bright light, it's too thoroughly seen, it stays inert. This has happened to me on occasions when I've got caught up in research, learnt more than I need about jellyfish swarming behaviour or fibre optic cables or whatever it is, and the subject has lain there, lifeless, exposed, bloated with information. Then I must take it into the dark, give my eyes a rest and use my other senses to get to what really matters.

I once took part in a conference where I was asked to talk about the writing process. I tried to describe it, calling on the metaphor of the dark space in which meaning is groped for, and can sometimes be touched and captured. An academic in another discipline responded irritably by saying 'It's a hundred and fifty years since Keats and his negative capability – I do think poets should have got past all that by now.' I was blindsided. It was as if he thought creativity and our ways of approaching it were stamped with 'use by' dates.

Of course he was right to mention Keats, that high Romantic, who coined the term 'negative capability' and whose 'Ode to a Nightingale' is its apotheosis. 'I cannot see what flowers are at my feet,' he writes, but it's not a complaint: in the dark he is returned to a state of uncertainty and receptivity, drawn into deeper insight by the scents and sounds of the night. My fellow panellist would have no truck with that. Rehearsing the scene several years on (as I do), I imagine him passing the microphone back to me (which he didn't) and I am neither winded nor outraged but gifted with sudden eloquence, and I say something stately like: It may be a very old idea, and it can indeed be problematic, because when soft-focus notions of *feeling* are allowed to overflow and wash away everything else we end up with a blur of sentimentality, but really Keats was just giving a name to something artists and writers of different persuasions have recognised, in all times and traditions, which is that rational thought – no matter how industrious they might be in applying it – is only one way of encountering truth, and that it's when they relax their grip on the empirical realities that other dimensions become apparent.

As far as anyone knows, Keats only used the term 'negative capability' once, which makes its longevity all the more significant. That single coining occurs in a letter to his brothers in 1817, where he proposes that the writer ought to be 'capable of being in uncertainties, mysteries, doubts, without any irritable reaching after fact and reason'. Other poets have gone further and talked about the need to relinquish control, or to 'get out of the way of the poem'. Some have spoken of a muse that guides, anoints or permits them; others of the writing coming through them, like a message spelled out on a Ouija board. Since Freud and Jung, many have conceived of the poem arising from the unconscious, and some have sought special access by writing under the influence of laudanum, or setting an alarm for two a.m. and composing a poem while half asleep. Workshop exercises, too, have grown up from this idea, like 'inkwasting', a form of deliberate distraction in which you write fast without thinking, sometimes carrying out some other mental task at the same time. Tutoring a group of intimidatingly good young poets some years ago, I used to get them to count to fifty out loud while writing on the subject of swimming, a headache, a train journey, or whatever. When they got too good at it, I made them count backwards. The idea is to catch yourself unawares, so that when you read back what you've written you're surprised by it. It certainly won't be a poem, though there might be a line or a phrase that eventually finds its way into a poem. It's a warm-up exercise, a way of getting yourself out of the glare of carefulness into the shadowy realm of intuition.

In that realm you rely not on night vision, but on dark vision. The aim is not to see *through* the dark, but to see the dark: to get past what is already manifest, and find your way into latency. And it brings with it a sense of incapacity which I find essential to writing. As I feel my way around in a space, unable to know the shape and size of it but mapping it somehow through touch, I'm back where I began. There are no certainties. I can't just fall back on skills and techniques acquired over the years of writing. I have the exhilarating feeling of starting afresh.

But so much for metaphor. Now I'm back in the actual night and looking around me at the real thing. Atomised city darkness, with streetlamps and skyglow and sirens and the scream of a fox. Or the interrogative darkness of the train window, when we go between towns and all I can see is my own wan reflection. Or else a warm summer night, on holiday somewhere, when I wander off from the candlelit table and the circle of talk and laughter, over the grass where my feet get wet with dew and the air pulses with the sound of crickets. I push open the door of the old barn and go in, brushing imaginary cobwebs from my face and hair, and grope my way to the splintery table I dragged across the gravel and installed here earlier in the day. I find the switch that turns on the Anglepoise lamp, and instantly the darkness at the open door is absolute. Moths dance in after me and cluster at the lamp. Most of what I thought I knew is dispelled by that vast, uncompromising dark, which contains more reality than all of it put together. I know that whatever I write here will have to be true.

Sick Room

'I have been one acquainted with the night,' wrote Robert Frost. The speaker of the poem is a walker, out in the dark and the rain, but in life the acquaintance can have a more domestic setting. Perhaps we are up at night to care for someone we love, bringing water, measuring medicine onto a spoon by lamplight, feeding a baby. Or perhaps insomnia or pain keep us from sleep, and we join a kind of subculture, inducted against our will into a knowledge of night's true character and extent. The hours between one and five in the morning are, by some sleight of hand, longer than the other twenty hours. You have run the whole orbit of your dreads and resolutions, yet when you turn over and look at the clock it's only 1.22. Your eyelids have forgotten how to close, so you stare into the indifferent dark and start again at the beginning. 1.36. Again. 1.50. How is this possible, when the daytime hours run and swirl away like water and there is never enough time for the things you have to do?

I think of it as a subculture because although there are many of us habitually awake at night our wakefulness goes unwitnessed by the majority, snoring or dreaming beside us and waking refreshed when the alarm goes off. If the slam of a car door or the scream of a fox woke them briefly at some point, they complain of having had a bad night. As bedtime approaches, their melatonin levels rise and they glide gracefully into drowsiness and from there into golden slumber, where they remain until morning. We see pictures of them on adverts for expensive mattresses, their sleeping faces blissfully pillowed, limbs relaxed between sumptuous cotton sheets. We move shuffling and yawning among these gilded creatures, thanking them for their sympathy and their helpful suggestions about herbal teas and warm baths.

In January 2020 I fell ill with meningitis. It was abrupt and decisive; I was felled. I had lain down on the bed in the afternoon to sleep off a headache, and when I woke up everything felt wrong. It was 3.40, and I foggily recalled that I was supposed to make an important phone call at 3.30. I crawled upstairs, dialled the number and tried to speak, but by now I had *become* the headache – I was possessed by it, it had seized control of speech and action, and what remained of my own will was squeezed so thin it was barely there at all. It was briefly interesting to find myself so thoroughly displaced within my own body. I laid the phone down carefully and stumbled back to bed. I wanted only one thing, and that was darkness. Since darkness is often characterised as nothingness, it may be that what I wanted was *nothing*.

Photophobia is one of the commonest and most well-known effects of meningitis. Inflammation in the lining of the brain, and a reaction in the trigeminal nerve, triggers a painful sensitivity to light. It's not a phobia in the usual sense – an irrational fear, or an anxiety disorder – but the aversion is so strong that it soon drags along with it a litany of neurotic thoughts. Please God don't let anyone come in here with a mobile phone. Don't let them turn the edge of the curtain to look out at the street. If they absent-mindedly press the light switch *it will kill me*.

The photophobia coexisted with a morbid fear of sleep,

which had turned into a swollen exaggeration of itself: a state so sudden and vertiginously steep I was afraid I wouldn't be able to haul myself back from it. I slept most of the time, and during the day it was tolerable because I could be checked on at regular intervals, but the sleep of night was a terrifying prospect. Sleep is often portrayed as adjacent to death – in Greek mythology, Hypnos and Thanatos are twins – and this adjacency was now terrifyingly real to me. Darkness was essential, and feverishly desired; at the same time it was a source of existential dread. Day and night ceased to exist. Dreams and hallucinations flowed together like two streams meeting and becoming one.

I was returned to a state of torpor I hadn't known since the ordinary illnesses of childhood. Whooping cough, measles, mumps, rubella, chickenpox: according to my mother I had them all in the space of a single year, though that seems incredible. But before the vaccines came along they were the inevitable diseases of childhood; we all had them. One of my earliest memories is of lying in bed with scarlet fever, waiting miserably all afternoon for my father to come home from work and call in to see me. When evening came at last he walked straight into the house and up the stairs without even taking his coat off. The curtains were drawn, and as he came into the room light from the landing shoved its way in behind him. He asked how I was feeling, and I turned over and said, 'Go away.' He went away, and the dark crept out from its hiding places and resumed its vigil.

The meningitis ran its course, and I began to recover. It left behind some tokens of its visit, like the vertigo, and the gaps in memory I visualised as sinkholes, and the way my eyes recoiled from words on a screen. Then there was chronic back pain which no one understood – nothing much to see on the

MRI, just nerves going off like a burglar alarm, all night every night. After a few weeks of this I was demented from lack of sleep; the anxiety I'd lived with since childhood burst out into the open, I was hungry and nauseous all the time, I wept at the breakfast table and pleaded with mirrors to let me get some rest.

Three a.m. It's at this small hour that most human deaths occur. The circadian cycle ebbs low at this time, the immune system relaxes, and the body, after lying in relative stillness for a while, makes fewer and less urgent calls on the brain. Blood flow slows, power is dimmed, the to and fro of communication subsides. In this later part of the night, we are at our most vulnerable.

Our minds know it, and this is an element in our fear of the dark. Thoughts of illness, mortality and death take on a different flavour at this time. Flexible ways of thinking that are habitual in the daytime abandon us at night; now context is stripped away, and balancing joys and other redemptive factors look remote or unreal. Lying awake in the dark can narrow our sense of reality to a single point of cold acuity, bleak and inescapable.

When I'm active in the day I am intensely engaged with my material surroundings: the people and objects I touch, the food I taste, the rain and sun on my skin. Sensation streams through me, wild and various and unceasing. But at three a.m. I have none of this. I have withdrawn from physical engagement with the world beyond the bed, the life of the senses that keeps me stitched in to the fabric of everything. I am detached and dislocated. The dark night of the body gives rise to the dark night of the soul, and my separation from the material world deepens it. This is one of the reasons why insomniacs are often advised to get up, go downstairs and put the kettle on. The simple actions of turning on the tap, getting a mug out of the cupboard, stepping on the cool tiles in my bare feet, can help remind me that I belong to the world.

When my back pain was at its worst, and it was impossible to lie in any position, I would put my boots on and go out into the garden or down the lane to the bridge. Tawny owls called across the valley, moving from wood to wood and mapping the space between them, one sending out its question *ke-wick ke-wick* and another replying with those long expressive hooting notes that carry so far in the dark. Touching the rough wood of the gate and feeling the weight of the latch, I felt a rush of relief: it was so good to be *actual*. As spring dawdled into summer there would sometimes be a grey ribbon of light on the eastern horizon at such times, maybe even the first bird calls of the coming day. Moving around was good for the pain, and being outdoors put it in perspective. Sensation streamed through me; I was part of something greater. My suffering was less total and less consequential than I'd thought. As the philosopher Erazim Kohák writes, 'When humans no longer think themselves the measure of all things, their pain is no longer a cosmic catastrophe. It becomes part of a greater whole.'

All the effects of the meningitis eased gradually over time. Five years on I still wake at intervals throughout the night, but I know I'll get back to sleep in the end. That makes me reluctant to get out of bed, and I look for another way to break the spell. I've learnt that cancelling the darkness is not the answer. If I reach out and snap on the bedside lamp, the bleakness doesn't retreat but, on the contrary, seems to spill from me and spread outwards to fill the room. I sit up and open a book, but I know it's just a distraction; when I raise my eyes from the page the room, the furniture, the objects around me still have that icy, staring quality. At times like that, the ordinariness of daytime is impossible to conjure with light alone.

But sound changes everything. I find solace lying in the dark with the radio propped on the pillow, close to my ear, and the volume low. It feels like a luxury and a relief to be joined to my fellow humans this way – at least to those acquainted with the night as I am.

Certain kinds of talk will do, if they are dull enough. But music is best. Darkness and music are the perfect combination. Without vision pulling relentlessly on my attention, without its insistent and enervating demands, I find it possible to listen more deeply and expansively, to have the fat and marrow of the music. After a while I'm not sure whether it is inside me or I am inside it, and either way it's so voluptuous I don't care. As I begin to soften into sleep I experience, for a brief slippery moment, a sense of edgelessness, as if I were unravelling gently into the dark and letting go of my place in time and space. Oddly enough, this is not so very different from the experience that felt unbearable in the silence earlier, but now it's warm and starry and consoling.

Bed

My best friend Sally owned a monstrous doll, the size of an actual real-life toddler. It could be made to walk across the room on its stiff legs, though it had a tendency to overbalance and fall sideways. Emerging from its back was a cord with a pink plastic ring on the end, and if you pulled it the doll would speak. Its vocabulary was limited to half a dozen words, all spoken in a robotic monotone, but of course it was miraculous that it could speak at all.

Sally was a girl who loved dolls, and her bedroom was crowded with them: some were propped on the pink fluffy cushion, and others lay on the pink candlewick bedspread staring glassily at the ceiling. I, on the other hand, have a lifelong dislike of them and all their kind – masks, puppets, waxwork models. It was one of the differences I first noticed between us, and I thought it was remarkable that we could be such good friends nonetheless.

The life-size doll had a split personality. In its way it was quite fun to play with for a little while: pulling the cord and making it do its zombie walk, arms outstretched as it cried for Mama, was a horrid novelty in the daytime. But once the lights were out, and the voices of Sally's eternally warring parents fell quiet at last, I lay as rigid as a doll myself, eyes wide open in the dark, ears pricked for the slightest sound: a small shift in the far corner of the room, the rustle of nylon, the faint click of a plastic eyelid. I'd had little exposure at that age to the tropes of horror stories and films, so I guess Freud was right: dolls are innately disturbing and *unheimlich*, because they resemble humans but are not humans. I recoiled from them at all times, but in the dark – when I couldn't see them and observe the limits of that resemblance – they became truly sinister.

The doll was only one of the unpredictable presences in that house. There were also Sally's parents, kind but unhappy people prone to spectacular rows that could explode at any moment. Sally had a kind of sixth sense or aura which warned her when to take cover, and she would bundle me out of the room, or behind a door or sofa, where we hid quietly until it was all over. I see the two of us now, crouching in the dark sanctuary of the broom cupboard, spying through the gap where the door is slightly ajar. Sally's mum, still in her coat and shoes, is weeping furiously. She seizes a packet from her shopping bag, tears it open, takes out a piece of raw liver which she flings across the kitchen – an arc of blood pursues it, beautiful against the pale wall – and it strikes Sally's dad on the neck with a loud slap.

Angry parents, ghastly doll: these were the known dangers. But fear of the dark was more often located in what you didn't know. You were carried in from the car to an aunt's house after a long journey, more than half asleep. With a tenderness that surprised you, your dad would shut the car door as softly as he could, and as you were borne through the front door and along the hall the adults spoke in whispers. Then your parents worked together, in that practised way they had, to unbutton your cardigan, unbuckle your shoes, pull back the blanket and slide you into bed.

Some children are capable of dozing through these procedures and resuming a deep sleep as soon as they're in bed, but

I would snap awake as soon as the door closed, suddenly conscious that I was in the wrong bed in a room I didn't know. No way of telling what objects might populate this room. Perhaps even other people . . . but no, I was sure I was alone, in a space of indeterminate size and shape, a space filled to the brim with darkness, as if it were a liquid, like soup poured into one of my aunt's Tupperware boxes and the lid pressed down to seal it shut with me inside.

Were we more permeable in childhood? It seems to me that the darkness of my surroundings could soak right through my skin and into my bloodstream. How unsettling it was to be put to sleep in another room when we moved house or went on holiday, and how alien the character of the dark, depending on the thickness of the curtains, objects on the windowsill or the way the street was lit outside.

It could even get right inside my dreams. I recall a week on a wooded holiday park with a playground and a path down to the beach. The days were a cheerful riot of swimming and climbing and fighting with other kids, but by night the place took on a completely different personality. Through the thin curtain at the chalet window the moon seemed nosey and menacing in a way it never did at home, and I was disturbed by the twisted shapes of branches close to the glass, twitching and tossing in the wind. I was so tired after the day's activity that I would fall asleep quickly in spite of my fears, but a few hours later I would wake up screaming and fall out of my bunk. I never remembered what the nightmares were about; I just screamed and fell on the wooden floor with a bang, once or twice a night, every night for a week.

When did the dark first cast its double spell of love and fear? I can't remember my earliest nights, when I must sometimes have lain awake in my cot in the box room at the back of the house. Our parents were not in favour of nightlights, and would not have left the hall light on and the door propped open. We were toughened up early. Many years later, my mother told me that I was put outside to sleep in my pram during the day, all through the legendary winter of 1962–3. It was the done thing at the time. There were eight-foot snowdrifts in places, and the sea froze a mile from shore, but I slept on, bundled up in blankets, hat and mittens. On those short afternoons, she would step out of the kitchen door when my nap was over and find it was already going dark. I picture her, closing the door behind her to keep in the warmth, and taking the narrow path Dad has shovelled and gritted that morning. She folds back the hood of the pram and there I am, lying awake with my eyes open wide on an altered world – flakes of snow against a navy-blue sky, a handful of stars.

Child psychologists say that babies are not afraid of the dark. Until around two years old they are indifferent to it; they may hate being left alone at night, but darkness itself does not frighten them. Perhaps it's only as we grow out of babyhood that we become aware of the powerlessness that is such a defining characteristic of childhood. In childhood, things happen to us, and we have no say in them. They're not necessarily bad things, but we can neither control nor predict them. Even in the habitual surroundings of home, darkness and light replaced one another in a pattern which I could only partly discern, and which changed not only with the seasons but also at the whim of my parents. They were the ones who had control of the light switch, and decided whether the door was open or shut. Once I was old enough to graduate from a cot to a bed, and tall enough to reach the switch, a cat-and-mouse game developed: when my dad put his head round the door to check on me I pretended to be asleep, and as soon as he'd gone back downstairs I put on the bedside lamp and picked up my book. He got wise to that, and I had to resort to a torch under the bedclothes. He became hypervigilant, checking repeatedly throughout the evening, craning his neck to peer up the stairs and detect a faint strip of light under my bedroom door. He even patrolled from the garden before he turned in, checking the square of my window for any sign of illicit reading. (He was himself a ravenous consumer of print in all its forms, from Dickens to *The Beezer*, so it wasn't really about the reading. I can only assume I was unbearable the next day if I hadn't slept enough.)

Darkness was already a powerful and paradoxical presence by then, and it had some strange and disturbing emanations. In the corner of my small bedroom under the eaves, there was a cupboard that housed the water tank. I judged it a privilege and an honour to share a room with this mysterious installation, and the cycle of weird noises it made after dark. It was a domestic god. The power it exerted over us all was generally benevolent, but I knew it possessed a moody potency that must be appeased. It was my job to placate and protect it, to allow it to speak in its strange language without interrupting, and to keep guard by sleeping beside it each night.

But there was another, more worrying aspect. My brothers and I had been warned repeatedly never to drink from the bathroom tap, in case there was a dead pigeon in the water tank, and one night, having lain awake ruminating on this awful idea, I found the courage to open the cupboard door, climb on a chair and investigate with a torch. If there was a dead pigeon I couldn't see it – all I could see was torchlight flashing on the black surface, and I toppled off the chair in terror, bruising my knee and shin.

I learnt my lesson. For weeks afterwards I stayed under the covers, no matter what. I don't know why the darkness under the bedclothes was safe, or why I was not afraid of the dark when I closed my eyes to sleep. When one night there was a tremendous bang like an explosion in the corner of the room, and my mum came running and switched on the light, I blinked theatrically, trying to pretend I'd slept through it. It turned out to be a toy ironing board which had collapsed with a crash. A toy ironing board! Whoever can have given me such a thing? I've loathed ironing ever since.

In infancy, I simply dwelt in the dark and neither loved nor feared it. Those responses came later, as I moved through childhood and adolescence. When I was growing up, darkness was a place of danger but also a place of self-discovery. It was a refuge too: when I felt exposed I looked to it for respite and protection.

As I reach back into memory, darkness seems to lean its weight on me, pressing me deeper inward. We all have our *heolstor* where we keep the difficult stuff, and there were certain incidents, long ago consigned to the past . . . nothing out of the ordinary, I guess, but formative in their own ways. If I go back and look at them, they might help me in my quest to understand the hold darkness has on me. Perhaps I'll work out how to separate those interwoven strands of love and fear, or perhaps I'll find the two are inextricable. Either way, I need to call on those early experiences. I summon them as witnesses, without whose testimony half the truth lies in deep shadow. They step forward into the present and address me. *You were here*, they say. *You did this. This happened to you.*

the headlights blazed a field gate then went out

You were to blame because you had accepted a lift home. He drove you most of the way and then veered off route. Though now you think about it 'veer' is the wrong word – he turned off quite calmly, indicating, checking the rearview mirror, taking the junction at a perfectly reasonable speed. *This is not the way*, you said, and at first you expected him to apologise, make a joke of it, do a U-turn. But he just kept driving and said nothing at all. *This is not the way*. It seemed to be the only thing you were capable of saying.

Next you were on a smaller road, and then a lane, and then a bumpy track. The headlights blazed a field gate then went out. A ticking sound as the engine cooled.

You were grabbing at the door and he was grabbing at your arm to stop you. You found the catch but the door slammed into a hedge and you couldn't get out. Meanwhile he kept making his modest demand: touch it, come on, just touch it. His voice was wheedling now and you almost felt sorry for him but then he forced your head towards him with his fist and you slammed and slammed and slammed the door against the hedge and he must have thought of the paintwork because he said in quite a different voice *You bitch* and revved up the car and hit reverse. You nearly fell out onto the lane and again you grabbed at the door but this time to shut it, which struck you as funny but you didn't laugh.

Outside the house he gave a dignified speech about how he'd decided to let you go as it wouldn't be fair on you because you were too young. Then you walked up the dark path and you let yourself in and went straight upstairs to your dark room and hoped he wouldn't remember where you lived.

Cosmic

And unable now to see your own hand in front of you,
you are actual size among your equals

under the twelve constellations
the five wild zones of heaven –

— Karen Solie, 'The Grasslands'

Landing

My childhood home holds so much darkness. The landing is particularly dense with it. Perhaps dark, like heat, always rises to the top of a house.

I slip out of bed when everyone else is asleep, tiptoeing out of my room and onto the small square of landing. I can cross it in five steps, but now I press close to the wall and stay clear of my brother's bedroom door, because it's a solitary expedition I want.

Tucked in under the eaves is a table with a plant standing on it: a mother of millions, the kind that propagates by producing tiny plantlets along the edges of its leaves. I don't really like to look at that plant, even in the daytime. Sometimes I count them, though I try not to. I can't see how there will ever be enough pots, enough space on the windowsills, enough neighbours willing to take them all in.

I know every inch of our house by day, but the night house is a different matter. I have noticed that after dark it undergoes subtle changes of shape: the staircase stretches longer and steeper, shadow pulls at corners and sharpens their angles, ceilings drift and float. I am an explorer, and it's my task to map these altered spaces where no one else goes.

Like most of the houses on this 1960s estate, ours is a dormer bungalow, with two small bedrooms in the roof space for my brother and me. The landing too belongs to us; our parents visit, and lately the baby has learnt to crawl upstairs, but they are only ever here on sufferance. If my brother has friends in his room, making pointless things out of Lego or Airfix, I keep my door shut to express disapproval. My friends and I prefer to be outdoors, or in faintly illicit places like the shed or the coal bunker.

I love the word 'dormer'. I have read the first few pages of *The House in Dormer Forest*, and the Elizabethan manor in its damp and melancholy valley is very real in my imagination, even though I have no real idea what an Elizabethan manor looks like. It also makes me think of the dormitory, an alluring place which is so completely outside my experience that I regard it as fictional. I don't think much of Enid Blyton's school stories, but I like the Jennings and Derbyshire books, sneaked from my brother's shelves. They have a rich vocabulary all their own: people are always 'skulking in the dorm' or 'beetling off to the libe', they have 'prep' to do in the evenings and their minds often 'boggle on all cylinders'. At my school there is no 'common room' where I can 'lie doggo', only the stationery cupboard, though that too is alluring in its own way, especially now I have been appointed to the dream job of paper monitor.

Of course, I never carry a torch on these night-time operations; whatever ambient light there is I use, but otherwise I rely on my sense of touch. No moon tonight, so I can't see the mother of millions. I think of running my finger along a leaf to feel the babies, but it's easy to detach them accidentally so I reach for the rough pot instead, and the old saucer it stands on. It's cold and featureless to the touch, but I know it intimately; it comes from a tea-set we must once have had, white with pink flowers, and it's the only survivor, apart from a cup with a broken handle that lives in the fridge and holds the bacon fat my dad saves from the frying pan on Saturdays.

The kitchen feels very far away. It will be a challenge to get there, and I will need to be brave, because the enterprise is fraught with danger. For instance, the fridge is a new one and has a small light inside that comes on when you open the door. It isn't much in the daytime, but at night it spills its yellow brilliance across the floor, loud and brash as a trumpet fanfare. I must remember not to open the fridge.

Now as I think of the distant rooms downstairs the landing seems to enlarge, its three walls yield and fall softly away. I am out in the open night, which is cool and infinite. I breathe the darkness. It clings to my hair and pyjamas. It grazes my face like spiderweb, and my hands and feet feel inky with it. I pause at the top of the staircase, and touch the wall very gently to guide and steady me on the way down. I shiver, wishing I hadn't thought of spiderwebs. Or of the mother of million babies breaking off and rolling onto the floor to die. Thoughts, too, grow and spread in the dark.

The wallpaper has a spongy texture you can dent with a fingernail. I trace its pattern and find the places where the edges meet. I would like to peel back one of those edges and feel what the wall is like under its paper coat, but I know I wouldn't get away with it. I feel the give of the stair carpet under my bare foot as I place it carefully, counting the stairs, knowing which are prone to creak, making myself weigh as little as I possibly can. Reaching the foot of the stairs, and turning the right-angle into the hall, is the moment of greatest hazard, because I must pass very close to the door of my parents' bedroom. One false step here and the game will be up. On the other side of the hall is the baby's door. Waking a baby is an unforgivable crime.

Exploration is an end in itself, but there is always a mission. It requires a visit to the pantry, whose door handle makes a distinctive squeak and must be treated with special care. The fuse-box is located here, and its low and mysterious buzzing fills the tiny space. The buzzing is a kind of thinking, the fuse-box the brain that goes on solving problems while everyone sleeps. It's much louder in the dark, and I worry that a light sleeper might be woken by it. But this was never meant to be easy. I know that at any second a switch could be flicked, the kitchen suddenly dazzling, and I in my red pyjamas exposed to the glare of my parents' accusations. As a precaution I have read up on sleepwalking and spent time at the mirror practising a glassy stare.

The return journey will be doubly risky because it's easy to get complacent. If all goes well, I'll be back in bed with my spoils and no one the wiser. I will probably have dipped my finger in the sugar bowl, the butter dish or both, but the real prize is always a single rectangular sheet of rice paper. It isn't the easiest thing to take, as the edges tend to crumble when handled and shed fragments on the floor. It requires skill and delicacy to bring it back to bed intact. It tastes of nothing at all, but I am entranced by it all the same. Paper you can eat! It's so white it glows in my hands. Each mouthful melts on the tongue and is almost nothing by the time I swallow it. One night I'll work out a way of writing on it first . . . and next thing I know it's morning, and daylight is flooding my room with ordinariness, and I run down the stairs without taking care at all, and my parents are getting the breakfast as if nothing has happened.

Lane

A house, whatever its age, is a provisional and temporary thing. I open the cottage door on a winter night, and instantly I'm engulfed by the black, rushing, torrential world outside, or by the electrifying proximity of the night sky. It reminds me of camping trips in wild places, when I would lie awake, shielded by a sheet of flapping nylon from the turbulent night, rocked by wind and rattled by hail, prowled around by unknown somethings.

Living at close quarters with the dark, I notice how wrong we get it. The clues are there in the language. It's ascribed spatial qualities (usually depth) and temperature (cold), though in truth it possesses neither; for a northern European, the warmth of night in parts of the southern hemisphere can be a revelation. We are said to be 'overtaken' by it, as if all day it had been pursuing us on horseback, and now galloped ahead leaving us weary and defeated. We routinely speak of it in an impoverished vocabulary with miserable connotations – it 'shrouds' things, it is 'gloomy' or 'murky'. The hours of darkness throng with life, but we insist on calling them 'the dead of night'.

Stepping outside the door tonight, I remark as I have so many times that it's 'pitch dark'. I'm referring to the proverbial pitch – the black tarry substance applied to boats to make them waterproof – but surely there has been some conceptual slippage, because when I utter the phrase I'm half-imagining being pitched into the dark: thrown headlong into that altered and disorientating state. When I checked my dictionary the other day, I found it gave as an example the sentence 'She pitched forward into blackness', which suggests that the two very different meanings have become thoroughly tangled. And there are other faint inflections in the word – pitch as level of intensity; pitch as the slope of a roof, steep and vertiginous; pitch as something fateful, like a pitched battle.

The true dark outside our cottage door may not be quite so ominous, but whatever the conditions – 'in the bare moonlight or the thick-furred gloom', as the poet Edward Thomas puts it – setting off up the lane is an expedition in itself. Still, if I want to pick up my messages it has to be done. I pull on coat and hiking boots, gloves and hat, and set off to walk to the stile at the top where there's good 4G. It's not far, and it's both a necessity and a small adventure. I made the decision many years ago, tramping the dunes and wild estuarial beaches of north-west England, day and night in every season, to designate myself an all-weather walker; I knew that otherwise I'd miss out on too much. That decision stands me in good stead here.

I don't like to carry a torch unless I really have to. It destroys night vision, obliterating everything beyond its small intense circle of light. I prefer to navigate by memory, as generations of workers and walkers must have done before me. Until recently the lane would have been nothing but stones and mud, and all but impassible after heavy rain. Nowadays there is tarmac, rough and potholed, with a broad strip of grass growing along the centre, a line that guides me as I walk, only dimly visible but felt underfoot. Navigation is a whole-body, multisensory task. Gullies either side carry the water, channelling it through mossy culverts of unknown vintage, underground to the river at the bottom of the valley. They frequently get clogged with fallen leaves and have to be dug out, but even then they are easily overwhelmed by the sheer volume of water, which gushes across the lane and courses downhill, dragging with it a cargo of branches and stones.

The twists and turns of the lane, the unevenness of the ground, the creaking trees and overhanging branches are all so familiar I borrow them as waymarks, and in the dark I call each to mind in turn, picturing it as I approach. If I'm alone I greet them along the way, using the private names I've given them, a litany that marks off each small stage of the journey and draws my surroundings close. Deer Crossing, Near Stile, Oak Hollow, Ivy Tods . . . I check them off in the right order, drawing and redrawing my mental map.

About halfway I come to the Knee Tree: a twisted old beech, facing east in a perpetual attitude of prayer. Its great smooth bended knee is discernible except on completely moonless nights, but the darkness aggregates at this point, where the lane swings round to the left, and the hedge line is breached by a ditch: all that's left of a gateway and a track that led to an old farm, now vanished. Our antecedents would have turned off here, would have driven cattle along the track, or carried a newborn lamb or a bale of hay through the gate and dragged it shut behind them. An ancient oak spreads extra shadow. I like to pause here, on the pretence of catching my breath. There's usually something rustling in the undergrowth, and always the sound of water, finding its own invisible route, naming its own familiars. This ghost junction, where one way used to meet another, is a place of crowding dark and whisperings.

The lane seems steeper by night, and narrower too, with

dips and holes sharp enough to twist an ankle. I visualise them, trying to remember where the broken sections are, where floodwater and debris have bitten chunks out of the edges and the way grows thin and scalloped. A little grey light hangs in the air ahead, as I push on towards Badger Gate and Top Stile. Not far from the road now, where by day there would be cars, but now there's stillness between the beech hedges, unless you're an owl scanning the ground for mice and diving for the kill. A country road at night is timeless; standing blinking into the invisible distance I half expect to hear the clip of approaching hooves and the rattle of cart wheels.

From here it's five miles to the nearest streetlamp, though whichever direction you take there are sources of illumination along the way: the windows of houses; the warm yellow light that spills, mingled with music and voices, from the open door of a pub; a security fitting with motion sensor outside the community shop. Between these landmarks – or lighthouses, as I think of them – you navigate by whatever the sky lends by way of celestial light. Sometimes car headlights strike you blind, and you stagger onto the verge and wait for your vision to recover. Your hearing is sharpened, and sound provides another set of bearings: the rattle of beech leaves in the wind, sheep coughing behind a hedge, the distant bark of a fox.

Electricity came late to the scattered settlements in these hills, and until it came the winter darkness was profound. Journeying over the rough lanes by foot or on horseback on a moonless night, you might pass only the occasional window glowing dimly with firelight. However well you knew the terrain, there was always a risk of losing your footing and ending up in a ditch. The thick mists and heavy rains we get here made it even harder to follow the twists and turns of the lanes that wind their way up and down the valleys. You probably had little to worry about in the way of crime in this sparsely populated place, though night has always provided the essential cover for transgressions of the law. No doubt as you passed a particular wood you might recall with a shudder the notorious murder (still now deeply engraved in local memory, over a hundred and fifty years later) of a tax collector who was taken there at night and bludgeoned to death with a pair of iron tongs. But crime in these parts has generally been petty: the poaching of ducks and rabbits from ponds and fields, petrol siphoning, fly-tipping.

I stand for a while and let my eyes adjust after the blue-lit screen of my phone. Beyond the crossroads, a light is on at one of the cottages, a building with an unusually chequered history. In the late nineteenth century it was used as a mission chapel. But further back, before the hymns and sermons, it was occupied by two brothers who made a good living by stealing sheep. Avoiding moonlit nights, and working without a lantern so that no neighbour would see them, they would creep through a gap in a hedge where they had seen the farmer put his flock to graze, take one or two and carry them home. The animals were then smuggled off to market, except for the occasional one slaughtered, salted down and hidden under the floorboards for the brothers' own consumption. Everyone knew everyone else round here, and nothing stayed secret for long. The two brothers were prime suspects for a long time before the police were eventually persuaded to investigate.

The Victorian missionaries who set up their chapels here had plenty to say about moral and spiritual darkness, but life was harsh up in the hills. Hunger and destitution were only a bad winter away, and in hard times people not usually given to thieving got desperate enough to trespass on neighbouring fields at night and dig up a few turnips. But the loss of a sheep was a different matter: it meant hardship for the subsistence farmer, already scraping a precarious living from the land. The penalty, too, was harsh: after trial and conviction the brothers were transported to Botany Bay, where their names disappear from the records. It's hard to imagine what kind of life they might have made for themselves, or how they adapted to the subtropical conditions, the searing light.

The window goes dark at the old chapel, and it's time to head home, striding confidently now, naming my waymarks in reverse – Knee Tree, Sheep Field, Deer Crossing – and here on the turn of the lane are the lights of our cottage, and the scent of woodsmoke on the air.

Abundant warmth, and food on the table – I'm grateful for the comforts that allow me to enjoy being out in the dark. But still. I feel in my coat pocket for the small charm I carry on my night walks, a smooth piece of chert with a hole through it. I'm told local farmers used to hang stones like these in their cowsheds to keep the cattle safe from harm till morning. They were intended not to chase away the darkness, but to protect it. Look: the dark eye is encircled by stone, its strength preserving the peace and shelter of the night while people and animals slept. My own charm is one I found on the streambed, fished out and claimed for its protective power. But I bet there's another like it, still hanging from the cobwebbed rafters of an old cowshed, not far from here.

Sky

Four o'clock, and the day is slipping away already. It's the end of a cloudy December afternoon, and now an ashy colour drifts in at the edge of the sky, like paint a shade or two deeper muddied in with the pale. Threads of ochre curl through it: I might take them for smoke, but our fire is not yet lit and no neighbour's smoke reaches us here.

The green of the winter grass is more saturated in the darkening afternoon, more intense in these last moments before it loses definition. I can see the tall birch, the chalky white of its trunk and the witches' broom clinging close to the top. The big old oak lost its leaves weeks ago, but its shaggy coat of ivy ripples in the breeze. The beech hedges suffuse the coming darkness with warmth and depth of colour – deep, reflective russet, plum brown, horse brown.

Whenever I can, I try to be outside at dusk, or owl-light. It's a good time to take a break from my own preoccupations and reconnect with what really matters: the rhythms of day and night, the rotation of the seasons, the movements and sounds of birds and animals as the earth turns and my small patch of it moves into shadow. The cottage itself is comfortable with darkness, lets it flow gently into the spaces between gate and wall and path, right up to the doors and windows as it has always done. Rooks row overhead to their roosts beyond the river. A pheasant strolls past, muttering to itself. The pond shivers, and brown leaves on the surface shift and settle.

Darkness is said to 'fall' – perhaps like a stage curtain falling at the end of a play, cutting off the audience from the scene they have just witnessed, the drama played out and the actors gone, the doors now standing perilously open onto corridors and stairways – but it's demonstrably untrue. When I'm out at dusk, I see it materialise first on the ground, condensing in the dips and hollows, then rise gradually, softening the outlines of the trees and the old bridge, seeping up like liquid and filling the valley, while the sky is still pale and glassy.

They come in softly, tactfully, those deeper shades in the sky. They take their time. The earth turns smoothly in space, sharing out the light. Sharing the dark, too – the wild, restorative dark.

In winter the nights feel not only long, but total. Darkness wells up sooner than I expect – standing at the kitchen sink I glance at the window and am surprised to see my own reflection there already – and rises steadily until we are underwater.

Of course it's not total. There are many much darker places: deserts, mountains, forests, places so far from human habitation that they are not touched by artificial light. Their pristine darkness is threatened, however, not just by light pollution from earth but also by fleets of satellites, which are coated with reflective metal plating designed to protect the electronic equipment inside. Without international agreements on space governance to curb them, they are multiplying rapidly. Each is a dazzling distraction to the eye, and together they are starting to affect overall levels of light, and damage the view of the cosmos for professional astronomer and ordinary stargazer alike.

Astronomers use the Bortle scale, established twenty years ago, to measure the quality of the night sky and identify the best sites for observation. The scale runs from 9 to 1, with Class 9 typical of the inner city and Class 1 a rural sky so dark that the Milky Way casts shadows on the ground. I have never seen such shadows here. On a good night this part of the hills might rank as 2, representing a 'truly dark site'. In less favourable conditions it would be 3, when a watery glow touches the northern edge of the sky – a trace of light pollution from the nearest town, reflected by low cloud, pressed close to the horizon by the weight of the dark.

Jo, an astronomer and dark sky educator, has brought a telescope and set it up on our front path. It's her first time in the Blackdowns, and she's keen to see how conditions here compare with those on Exmoor an hour's drive away. There has been a run of wet weather this week, and I've been up and down the lane with my phone, hitting refresh on the Met Office forecast. When she arrives it's overcast, but later the cloud is predicted to break and conditions improve, though the timing is uncertain. She pops out every few minutes to see what's happening, while we wait indoors, lingering over our red wine and watching the clock.

At about nine she calls us to come and look, and we pile on coats and boots and go out. The sky is a patchwork of cloudy and starry pieces – no, not a patchwork, because now I see that it's alive, and shifting moment by moment as the cloud moves and clears. To the east I can see the thrown milk of our own galaxy, or rather an edge-on view of it from where I'm looking, because our entire solar system is just a tiny dot on one of the arms of the spiral.

A few familiar constellations emerge: the Great Bear, Orion, the Pleiades or Seven Sisters. Jo tells me that there are actually over a thousand Sisters, and that a sketch made by Galileo in 1610 shows thirty-six of them, seen through his telescope. Tonight, when I put my eye to the lens, I count fifteen.

The cloud thins and disperses, and the sky grows deeper, not only in colour but spatially too, opening up into three dimensions so that now I can see stars beyond stars beyond stars, in a space without edges. I remember something C. S. Lewis said about medieval cosmology: that looking at the night sky through modern eyes is like looking out at a view that fades away into endless distance, whereas through medieval eyes it was like looking up at a towering building. It would be impossible for me to believe in the towering building, it's so hard to imagine ourselves into other ways of seeing. But I'm not sure I can really believe in a space without edges either – it's as if my mind doesn't quite know how to make sense of such a concept.

Meanwhile Jo is explaining that until a hundred years ago it was thought that the Milky Way might be the whole universe. Edwin Hubble's announcement that the 'nebula' Andromeda was actually a galaxy could hardly have been more momentous. At a stroke it expanded our view of the universe, and shrank our sense of importance, since now our own galaxy was just one of many. The idea had been around for a long time, and as early as 1755 Immanuel Kant – not a scientist but a philosopher – was suggesting that the dim fuzzy patches called nebulae were actually 'island universes', or galaxies like our own – but it took a hundred and fifty years to prove it. Now we know that there are more stars in the universe than there are grains of sand in the world.

Andromeda is visible to the naked eye tonight, as a faint smudge like a thumbprint on a window at night. When I pick up the binoculars, the view is so wobbly and disorientating

I can't find it, until I work out that I can rest them on the wall and crouch to look through them, and then I see its distinctive elliptical shape quite easily. When I use the telescope, there's a white-hot blob at the centre of the ellipse. That blob is the core of the galaxy, and is made up of billions of stars, whose light, says Jo, has taken two and a half million years to reach us.

My head is starting to reel with all these impossible numbers, and it's a relief to have my hands on solid and tangible things: the familiar heft of the binoculars and the flat bit of wall to rest them on. The binoculars are my dad's old ones, large and robust, made in the Soviet Union in the 1970s; I recall them slung around his neck on hillwalking expeditions. It makes me happy to be using them tonight, and I'm surprised by how powerfully they change the view. I can see not only Jupiter but three dots that must be its moons! I'm feeling pretty good about this achievement until Jo tells me there are ninety-five of them.

Orion nebula, Beehive Cluster, Betelgeuse and Bellatrix, Aldebaran – there comes a point in the evening when I lose track of them and know I won't remember what I've seen. Names, numbers, distances: right now I want to make them all transparent so that I can see straight through them to the stars themselves. I just want to look. The sight of all that unimaginable space fills me with awe, making me aware of my own smallness and the shortness of my mortal life – it's a cliché, really, but tonight there's an immediacy and an urgency to that feeling. Awe is a compound, and one of its ingredients is a kind of terror that can strike you to the heart. This is the gift of fear, I think: it returns me to reality. The lens screwed to a sharp focus, springing everything into proportion.

When I was in eastern Turkey recently, visiting the

prehistoric structures at Gobekli Tepe, I read a theory that some of the carvings, particularly those on Pillar 43 with its strange vultures, scorpion and headless man, are a kind of star map. The carvings are deeply mysterious, and there are many speculations about their meaning, but in this case the archaeologists working on the site were unconvinced. The carvings were made twelve thousand years ago, they said, and celestial objects would not have been grouped in the same way as they are in the Western world today. Similar controversy has surrounded claims that some of the Neolithic cave paintings at Lascaux may have been created as star maps, perhaps indicating seasonal change and hunting times. The temporal distance is so great, and the people who made these artefacts so remote, we can never really know how they constructed reality; but what must be beyond doubt is that people have always looked at the night sky and wanted to understand it. In the world of the hunter-gatherers who painted cave walls or carved elaborate pillars of stone, the starry sky was not optional or occasional as it is for most of us, but overwhelmingly present, and bright enough to break the Bortle Scale.

Tomorrow morning, while I'm filling the kettle, making toast, brushing my teeth, I will still be thinking about the stars I'm seeing now. I'll look out of the bathroom window and know they are still there, even if the sky is blue and empty apart from a few clouds. I will feel that knowledge in my bones: that for all our shiny satellites the earth is just a speck of dust in the dot of our own solar system, at the edge of our galaxy, which is one among trillions. Those colossal realities, which generally get lost as I go about my usual routines, will accompany me all day.

I will put away the binoculars in their leather case, which smells and feels like childhood, along with the two sets of coloured filters and the original cloth for cleaning the lenses. They're kept on an upstairs windowsill and mostly used to look at birds, but it turns out that they are excellent for stargazing too. As I look across these unimaginable distances tonight, I think of my parents, and wonder where if anywhere they are now. But perhaps *where* is all wrong, I think. Could I believe in a heaven that is not a place at all, but a time? For an instant, standing looking through those old lenses, I escape the confines of the here and now, and it feels possible, in this moment, to imagine that death is a matter of slipping out of this time into a parallel one – one that runs on another kind of clock. For a few seconds I see it, that pale smudge of possibility, before a small cloud crawls over and hides it again.

Cave, 1984

We are told that complete darkness does not exist anywhere on earth. Light, like water, will leak in through the smallest gap, and flow into whatever space it finds. And in places untouched by the sun, there are organisms that can generate their own light.

Our human perception convinces us otherwise. There's so much we cannot see; even with our night vision, we encounter in some spaces a darkness that seems unequivocal. The intense dark of the cave is like that.

In my early twenties I worked for a long, unforgettable summer at an outdoor pursuits centre in the Black Mountains in Wales. Things were different back then, and the safety regime was very relaxed. You turned up with no qualifications or background checks and were put in charge of a group of teenagers as they learnt mountaineering, rock climbing, kayaking, archery and rifle shooting. There was always an experienced instructor on each activity, but the rest of us were thrown in to sink or swim, sometimes literally. One of my earliest and most bruising experiences was on a three-day kayak expedition. I had once helped paddle a canoe – on a wild white river in Michigan – but this time I was in a single kayak so it was all down to me. I spent most of the first day capsizing; the bag of clothes, food and camping gear tucked under my feet floated downriver and filled up with water before being fished out.

There was an understanding that each of us could opt out of one type of activity, if we had a really strong aversion to it. I was fit and ready for most kinds of assignment, but I made it clear on arrival that I did not want to go caving or potholing. I suffered from claustrophobia, I said. I mentioned a primary school trip to Peak Cavern at Castleton, where I'd had a panic attack. There was a derisive snort from the camp manager as he made a note. I do not think those notes were ever filed.

Inevitably the day came when a group leader was ill, and I was called in to cover for her on a day's caving at Porth yr Ogof. The instructor, known in accordance with our usual tradition as Phil Cave, was held in high esteem because he had an actual certificate of some sort, and was known as a tireless evangelist for his sport. This was worrying, because I hoped to be underground for as short a time as possible and had no wish to be converted. To the enthusiast, reluctance can be an irresistible challenge.

I'm happy to make brief descents, taking an uneven staircase into a dungeon, cellar or crypt, dimly lit so that I have to feel my way and strain to see the ancient inscription, or the prisoner's calendar scratched on the wall. I fondly remember a visit to a champagne cellar in Reims, where the monks had cut the shape of the staircase into the ceiling above the real staircase, making a three-dimensional plan so that they could reach up and feel their way down in the dark with a bottle in one hand. On such tourist visits I haven't gone very deep; my body has kept count of the steps, and is confident it can find its way back whenever it needs to. But this was to be a day expedition. I would be helplessly reliant on Phil, who enjoyed regaling us all in the pub with tales of rockfalls and flash floods and near-death experiences. We would be under-ground for hours, and I would not have an escape route.

Sometimes there is only one thing that can drive out fear, and that's fear. A member of our group, an athletic Dutch boy who had been the talk of the camp earlier that week for his reckless back-to-front abseiling technique, was unexpectedly ambushed by claustrophobia himself. We were about an hour into the expedition and had reached a narrow passage where we had first to crawl and then to squirm through a tight gap between protruding rocks. One or two of the kids up ahead were struggling; I could hear Phil urging them through. 'You ain't seen nothing yet,' he boomed cheerfully, and his words echoed portentously in that bare and dripping space. The Dutch boy halted on all fours halfway along the passage and began to scream.

Is there anything worse for someone only just controlling their own terror than someone screaming a few feet away? My throat and chest were bursting, I teetered on the brink of total panic . . . but how could I afford to fall in? There was no question of going back. I would have to find a way of suppress-ing my own fear in order to help him.

It was a long day. I coaxed that sobbing boy through narrow gaps and down vertiginous shafts, as he shrieked and begged and prayed in both our languages and declared we would not get out alive. He clawed at my face, grabbed at my hair and hung on as if saving himself from drowning. We were wearing head torches, and I remember trying to limit my movements so that the beam would not race and judder over surfaces, terrifying us both with its sudden exposures of the space we were in, the weird shapes jutting from the walls, the water leaking ominously down the rock. The boy became a second self; I soothed, encouraged and braced us both. If I hadn't been so afraid, I don't think I could have helped him at all; what would I have had to offer but my impatience? I think of him now as I write this and wonder how he remembers that day, which must have been both traumatic and humiliating. I wish I'd told him how grateful I was, because his fear was the saving of me.

Claustrophobia is defined as 'an irrational fear of confined places', but I see nothing irrational about it. In his book *Underland*, Robert Macfarlane describes it as 'the sharpest of all common phobias', observing that its symptoms can be triggered even by hearing stories of confinement. For some people, it's so extreme that it dictates where they can go and what they can do in their everyday lives. But for the rest of us – the many who are wary of lifts, and who sit sweating, hearts pounding, when a tube train stops too long between stations – it's not a disorder at all, but a normal response to the real possibility of becoming trapped. Lifts do sometimes break down between floors. Power failures are not unknown. Imagining scenarios like these, it can be hard to separate out the claustrophobia from the nyctophobia, or fear of the dark, because when the power goes off, so do the lights.

Those who can fight the fear and conquer it are courageous souls, and I envy them. But to lack that fear in the first place can only be a failure of imagination. This is one of the things I gabbled at that terrified boy: that he possessed a great imagination, which was a kind of superpower, but could be inconvenient at times. (We even managed to laugh, the two of us, over that feeble word 'inconvenient'.)

In the confined spaces of underground chambers and tunnels we are out of our natural element, and as Phil Cave told it in the pub the history of speleology is littered with the corpses of those who never made it back to the surface. He was eloquent, too, on the extremes to which cavers have been prepared to go in pursuit of exploration and discovery. One of his heroes was a French explorer called Michel Siffre, who had spent sixty-three days alone in a cave. Oh my god, sixty-three days? We were duly amazed. Didn't it send him loopy? 'Nope,' said Phil, pausing to take a long drag on his Marlboro. 'In fact he did it again, but this time for six months.'

Eventually the boy and I caught up with the rest of the group at our destination, a cavern where there was a large subterranean pool. They had been there for ages, eating their sandwiches and throwing stones into the water. The two of us sat down cautiously, with a clanking of karabiners, and I could hear him shuddering and sniffing beside me. After a while, Phil Cave instructed us all to switch off our torches and sit in silence for ten minutes. Even my boy, relieved to be out of those narrow passageways at last, was now calm enough to comply.

And there, suddenly, was the great discovery of the day. As soon as we flicked the switches we were rid of those nervous torch beams with their unpredictable jumping and dazzling. We were neither peering anxiously ahead nor squinting back at the way we'd come. We were not scanning the roof for signs of weakness or checking each other's faces for hints of fear. All of it fell away in a deep, embracing darkness that was like balm. I felt my claustrophobia soften and melt to nothing. The last giggle, the last cocky remark died away too, and silence stole over us.

Not really silence, of course. My other senses sharpened to compensate for the loss of vision. One small sound led to another, like a line in a poem disclosing the next. I could hear my own breathing, and the slight rustle of my clothing as I breathed. Distant drips of water, like tiny glass footsteps.

Even, perhaps, minute shifts in the surface of the pool, though what breeze could possibly be touching it? I remembered Phil and a couple of other instructors talking in the pub about a plan to go windsurfing here, and I sat puzzling over that before deciding they must have been having us on.

Then the liquid darkness rolled in again and I slipped free, thought of nothing, gave myself over to feeling. There was no longer any way of judging the size or shape of the cavern, and I felt the space flex and enlarge around me as the minutes passed, the walls first creeping outwards and then picking up speed and racing apart until spatial dimensions became meaningless. No sooner had I caught hold of this slippery perception than Phil was announcing the ten minutes was up and the torches were being flicked back on and the moment was lost.

It was the briefest taste of transcendence, a psychedelic experience in which the boundaries of my body lost their distinction and my sense of self loosened and dissolved. Those few minutes of sheer darkness revealed a blazing truth that I have remembered and occasionally recaptured in the years since: that in spite of all my everyday assumptions there is no division between me and the universe around me.

fluorescent lights that flickered when the bulb was going

You were ferociously self-conscious in those days. You looked for places where the light was attenuated and you were less exposed. The shadow thrown by a garden hedge, a gravel path between garages, a corner, a cloakroom.

You were an object to be viewed and scrutinised by others. They would admonish you for the way you looked – *Cheer up, give us a smile*. They could touch you if they wanted, that was understood. But even without touching there were other kinds of violence. The street was violent, the bus was violent, school was unutterably violent.

You were suddenly transferred to a new school. On the first day there was a crowd at the gate and the January light was merciless. There hadn't been time to buy the right uniform and your skirt was a mockery. The deputy head pretended they were expecting you. He walked you down a dazzling corridor, halting at each classroom door to point at the square of wired glass and say, 'Do you recognise anyone?' By the third door he was getting impatient so you said yes. 'In you go then,' he said. The noise was deafening. Sunlight surged in at the big windows.

At dinner a group of boys jumped the queue to investigate whether you had tits yet. For some reason you had been issued with the wrong face and the wrong body. The hall had no corners. There were fluorescent lights that flickered when the bulb was going. Tiny red spiders were running over the tables and people were squashing them with the back of a spoon.

A boy came out of technical drawing and hit you with a lovely wooden implement called a French curve. It made your arm swell up, and you were a spectacle. You were shown to the teacher, who let you sit in his stock cupboard to recover. It was dim in there but you noticed a kettle and an overflowing ashtray, and you realised you were not the only one who needed to hide from the light. You sat with your arm in a sling and were happy for the first time all year.

Power cut

Keep writing in the dark:
A record of the night, or
Words that pulled you from the depths of unknowing

— Denise Levertov, 'Writing in the Dark'

Cul-De-Sac

Last winter a storm brought down power lines and they stayed down for five days. It brought down trees, too, blocking the lane and making it impossible to drive out of the valley. The phone went dead, and we were cut off from everyone, except our neighbour's father who had the misfortune to be house-sitting for the weekend. We had a well full of water but no electricity to pump it to the house, so this was the moment to open the ten-litre bottle we keep for just such an emergency. There were plenty of logs in the shed and the woodburning stove kept us warm. We tied a fork to a stick and sat by the stove toasting bread. We propped a metal rack over a couple of flattish bits of wood and heated pans of soup and baked beans. Knowing it would be dark around four in the afternoon, we had to plan ahead and make sure candles and matches were placed in strategic locations and our one remaining working torch was ready to hand.

We were grateful for our home comforts, and for a while it was quite enjoyable. It brought back long-lost memories of the early 1970s, when there was a series of disputes over pay and conditions for coal miners, resulting in fuel short-ages, long power cuts and the three-day week. It must have been an anxious and gruelling time for our parents; daily life became harder, livelihoods were threatened and no one knew how long it would be before things returned to normal. For school-age children, though, it was blissful. Schools could only be kept open for a few hours a day, so we were sent home at lunchtime with a handful of worksheets to complete. In those pre-internet days there was no way of continuing lessons remotely, so we didn't have to sit through virtual classes as young people did during the Covid epidemic. Instead, I would head up the road to Sally's house, usually but not always remembering to take my worksheets with me, and we would sit on the sofa, snuggled up under a blanket, eating Wagon Wheels and taking it in turns to read aloud from *Jackie*.

The worksheets would get some cursory attention – just enough to satisfy teachers and parents that we had done *something* educational during those unsupervised hours – but once we'd scribbled a few sentences of comprehension and dashed off a page of sums we regarded ourselves as free to do whatever we liked until her mum got home from work. And even the sums and the comprehension were more fun done together with a glass of strawberry Nesquik to hand. There was a choice of three flavours in the cupboard at Sally's house, and stripy paper straws to drink it through. It got dark early those winter afternoons, and in my memory I see the two of us, slightly giddy with sugar, giggling over an eleven-plus practice paper in the dusk, aided by Sally's fancy colour-change torch. *Imagine you are a firework and tell a funny story about Bonfire Night.*

There's an elegiac quality to the memory, because Sally, always more grown-up than I was, got suddenly tall the next spring, started wearing mascara and was followed around by irritating boys, and our friendship began to wane. (Her twelfth birthday party was a last fling: a sophisticated affair that went on after dark, and culminated in a conga line that made its way around the block, pausing scandalously at the postbox on the corner for her to take off her knickers and post them.)

The cuts of 1972 went on for weeks and were planned in advance: a schedule was published in the papers so that you knew when the power would go off. But there were other, unexpected blackouts too, and they created a different kind of excitement, inserting fun and novelty into the ordinariness of a weekday evening. The thrill of that sudden flip into darkness, the television image imploding into a small bright star, the brief suspended moment, like a catch in the breath, before someone said, 'Not again!' Then the sound of my dad fumbling his way out of the room and into the pantry, bumping his arm on a doorframe and cursing under his breath, and the loud *thunk* as he tested the trip switch. The exciting smell of meths, and the wavering light of the hurricane lamp carried into the room and placed on the table. The biscuit tin brought out to keep everyone's spirits up. Perhaps a game of Sorry or Chinese chequers; dull games made much more interesting by lamplight. It seemed to me that we laughed more extravagantly, played more recklessly, bore our losses without the usual rancour. Later, the three of us would be allowed to see our way to bed with the battery-operated red plastic lanterns we'd been given for Christmas, though we had to turn them off as soon as we were in bed to save the batteries.

Sometimes we would go outside to see who else was affected, and soon all the neighbours would be there, and a strangely festive atmosphere would develop. The adults would stand around in the road, talking about the half-cooked dinner going cold in the oven and the spare blankets they could lend if anyone was worried about being cold in bed, while we ran about shining torches in each other's eyes and pretending to see ghosts in the cul-de-sac. I loved seeing all the houses in our road with their windows dark, and the streetlamps off too. I didn't know anything back then about light pollution. I had seen moths at the streetlamp on the corner, and I was sad when the boy next door told me they thought it was the moon. But if you were lucky on the night of a power cut, there would be stars, and a feeling of going back in time, to a world before everyone had boring light switches stuck to the wall and sat in their own houses in the evening watching the same things on their identical television sets. You were never scared of the dark at these times, because power cuts brought people together.

Campus

In Jill Tomlinson's classic picture book *The Owl Who Was Afraid of the Dark*, the baby owl – surrogate of the child reading the book – is assured that dark is not nasty, as he imagines it, but kind, fun, exciting and beautiful. In ascribing it positive attributes, the book aims to challenge habitual attitudes to the dark and get the reader thinking and feeling differently. It's a welcome piece of quiet subversion.

But sometimes, in an attempt to comfort and fortify the frightened child, an adult will deny the power of darkness altogether. When we say to a child that there is *nothing* scary about the dark, the child surely knows that this is preposterous.

Fear of the dark is commonly associated with childhood, as if it were something the child would or should outgrow, like skipping, tantrums or cuddly toys. In truth, what makes the difference is that we gradually acquire more autonomy as we grow into adults, and have a degree of choice over how much time to spend with the dark. We can switch lights on and keep them on as long as we like; we can take the most well-lit route home at night; we may even be able to factor this choice into our decisions about where to live and work. We can't eliminate darkness from our lives, but we can limit our exposure to it in ways that were not available to us when we were children.

Childhood is sometimes characterised as a kind of Dark

Ages, from which we emerge into the Enlightenment of adulthood. As we grow into adults we turn more readily to reason to guide us, reminding ourselves that our planet orbits the sun, rotating as it goes, moving in and out of the light. These fundamental facts reshape our thinking, and life experience does modify our expectations, but rather than melting away, our fears shift and regroup. We stop believing in monsters under the bed, and start worrying about human monsters who want to harm us.

During my first term at university, I was crossing campus one day when I was stopped in my tracks by the sight of a poster bearing the slogan RECLAIM THE NIGHT. I stood and looked, and felt something click into place in my mind as I understood for the first time that this was a collective as well as an individual struggle. There it was, in those three words: the extent of exclusion, how effectively whole communities or sections of society could be cut off from an entire realm of experience. That campaign opened my eyes to the conditioning I hadn't really noticed before. We girls had internalised a whole set of constraints, I now realised: we should go out in groups rather than alone, wear certain clothes instead of others, walk purposefully rather than stroll comfortably at a pace of our choosing. We had picked up rudimentary techniques for self-defence: a bunch of keys clutched in the fist would provide you with something to jab in an attacker's face; an old credit card cut in half was as sharp as a knife but did not count as an offensive weapon. We shouldn't have to carry this stuff, it was a mark of our dispossession. This rain-streaked poster tacked to a tree spoke to me of the thrilling possibility that if we got together we could do something about it.

It was the urban night we wanted to reclaim, and it was not darkness itself that dispossessed us of it but the threat of violence it could conceal. Nevertheless, most of us don't stop feeling scared of the dark when we venture into the country-side; indeed, for some people the fear increases. When my grandparents moved from the town to the edge of a village, they complained that they couldn't sleep at night because there were no streetlamps. Rural dark of the kind I live with now can be more intense than the city dweller expects.

One response to fear is to try to banish the thing you are afraid of. You can install floodlights and spotlights, rig up security lights with sensors that trigger every time a cat walks past. Another response is to try to destroy the fear itself, by denying and disavowing it, by attempting to force it out or vanquish it. But fear of the dark is neither a weakness nor an embarrassing hangover from childhood. It can't be stamped out like some individual foible, because it's an integral part of being human. 'After thousands of years,' writes Annie Dillard, 'we're still strangers to darkness, fearful aliens in an enemy camp.' In our remote past it was crucial for humans to be on high alert at night, to protect ourselves from predators; the association between darkness and danger is primeval, and we can't just wish it away.

I love the dark, and I am afraid of the dark, and I'm begin-ning to understand that this is not a contradiction. Our bodies have evolved to create this fear; there are physical mechanisms that generate it. Neuroscience shows that par-ticular parts of the brain are acutely affected by light and dark, notably the amygdala, which plays a primary role in process-ing anxiety. Light can make the anxious person feel better, at least temporarily: it suppresses activity in the amygdala and is associated with mood enhancement and 'fear extinction'. At the same time, light has a suppressive effect in another structure, the habenula, which among other functions plays a role in addiction. Decreased activity in the habenula is associated with greater expectation of reward, and this increased sensitivity encourages us to keep seeking the feel-good factor of light, even when it's bad for us. It's natural to seek the cheer and reassurance of light – the night traveller has always been drawn to the lit window, with its promise of safety and comfort – but artificial light is like sugar: the more we have the more we want.

As a lover of the dark, I want to think there's a better way. Rather than trying to eradicate fear, I wonder, might it be possible to think of it as a gift? Instead of trying to laugh it off or soothe it away, we might reflect that we have reason to be grateful for it. One of the characteristics of being human is that we have poor night vision, and that's bound to make us less comfortable in the dark. Even if there are no predators about, the risk of accident is higher: there are invisible obstacles, things to trip over and fall into. The discomfort we feel when we venture out at night makes us sharpen our awareness, prepare to protect ourselves if we need to.

A little while ago, I resolved to stop battling my fear of the dark, and to make peace with it instead. At the time I had persistent back pain, and I came to feel that the two phenomena had a lot in common. My physiotherapist taught me that pain is created by the brain as a way of drawing attention to injury, and it occurred to me that fear of the dark might work in a similar way: keeping me safe by alerting me to risk, prompting me to take extra care as I move around in the world. Neither pain nor fear is a perfect system, however. As my physio explained, the nerves in my back will sometimes fire even though there's nothing wrong with my spine. Likewise, I realised, my heart will sometimes trip and my skin prickle with fear even though there is no danger lying in wait.

It seems pointless to try and abolish either fear or pain; better to try and be grateful for both, and to accept that from time to time they will be triggered unnecessarily, like an alarm that can't tell the difference between a burglar and a gust of wind.

That's the theory, but in practice it isn't easy. I'm so vulnerable out here in the dark, with my weak night vision, my soft body, my insistence on walking on two legs. And looking up at the night sky I am confronted with the true scale of my vulnerability. I want to regard myself as powerful, but now I see how insignificant I am, how helplessly stranded in unimaginable space. The fear that courses through me then is of a totally different kind: pure existential terror that makes the usual anxieties feel like child's play.

Cave, 1962

'A new idea sprang up: I would not take a watch down with me. I would study, in the continual subterranean darkness, my loss of all notion of time.'

To be honest, I thought that Phil Cave was making up the stuff about Michel Siffre – or at least exaggerating. He did love an audience. But no, it was true. After reminiscing about that day trip to the cave in Wales, and recalling the way it changed my perception of space and time, I did a quick online search and there he was: much more famous than I'd expected, still something of a cult figure. They called him the Cave Man, and there were pictures of him in his twenties, his thirties, his sixties: kitted out for adventure, roped up for the descent or grinning as he emerged from some hole in the ground.

I bought my own copy of the battered paperback Phil had brandished in the pub that night: *Beyond Time*, Siffre's own account of his first solo expedition in 1962. As it happened I was about to go on an expedition of my own, to the Amazonian rainforest, and I tucked it into my rucksack with three or four other assorted books, chosen for their compactness rather than anything else. The rucksack went with me on the flight, on the minibus, on the truck and on the motorboat that took us three hours upriver to the forest lodge where we would stay. In our room, which was open on one side to the warm, pulsating dark, I took out the book, lay under the mosquito net and began reading by torchlight. It was the most incongruous subject matter, but I was quickly engrossed by the story of his descent into the cave and his description of that otherworld of rock and ice and raw darkness. The outermost corners of the room, already remote and uncertain, seemed to dissolve entirely as I turned the pages.

The story begins when he is a geology student and self-proclaimed enfant terrible. He's planning to spend time alone in the Scarasson abyss, a cave high in the Alps, close to the Italian border. It will be his second time in that cave; on his first trip he made the exciting discovery of a subterranean glacier, and now he's determined to go back and investigate it further. He hatches a plan to take a tent down with him, pitch it on the ice and stay for an extended period so that he can study the glacier in detail.

To begin with he envisages camping on the ice for fifteen days. This is not a modest ambition; there will be a host of practical problems to overcome. In a high altitude cave the air is unvaryingly cold, with a hundred per cent humidity, and he'll need the right clothing. Camping on a glacier will require special equipment, and all of this will cost money. He begins to sketch out the work he intends to do. Really, he tells himself, once he's worked out a way to pay for the gear he needs, it would make sense to stay longer; there's plenty more he could do while he's there. He decides to extend it to two months. This will mean setting a new record for the longest time spent camping underground, which is a pleasing bonus.

At that moment, the extraordinary new thought strikes him. Underground for sixty days, isolated from the cycle of day and night, what will happen to his sense of time? Will he lose the ability to track it, or will some internal mechanism continue to operate even in the absence of daylight?

What begins as a casual thought quickly turns into an obsession. An entirely new field of scientific enquiry opens up in his mind, and all but overwhelms the original purpose of the expedition. It was exciting to find the subterranean glacier, and he hoped to make new geological discoveries that would help him establish himself in his field. But this new ambition is on a different scale. Could he (while studying the glacier, of course) rediscover what he calls 'the original life rhythm of man'? Might he be able to demonstrate whether there really is such a thing as the body clock, and if so what makes it tick? 'In short,' he writes, 'I wanted to investigate time – that most inapprehensible and irreversible thing.'

It's easy to forget how young the science of chronobiology is. The cycle of night and day, and its influence on our bodies and minds, had long preoccupied biologists, and the idea of an internal or endogenous clock had been around for two hundred years or more. A famous experiment in 1729 proved that something of the sort existed in plants. The mimosa plant, like many others, opens its leaves during the day and closes them at night, and the experiment was designed to reveal what would happen to this regular pattern if the plant was not exposed to light. A mimosa was placed in a dark cupboard and the opening and closing of its leaves were observed to continue regardless. It was a simple finding, but its implications were huge: it meant that the mimosa must have some way of anticipating day and night, rather than simply reacting to it. Further experiments found that if the plant was kept in constant light its leaves opened one to two hours earlier than usual. The plant's innate awareness of time was not perfectly in synch with the actual time; in the absence of the usual light-dark cues it found its own rhythm.

Over time these same principles were extended from plants to humans. Julien-Joseph Virey, a naturalist and anthropologist, was the first to articulate it in terms of a clock. Writing at the turn of the nineteenth century, he asked: 'Doesn't this successive rotation of our functions, each day . . . establish a habitual and innate periodicity in the whole play of our organs? Is it not like a system of interlocking cogs, a kind of living clock, assembled by nature, driven by the rapid movement of the sun and our sphere?' The question is remarkable enough, but the really astonishing thing is the way he thought about its implications for health. When he applied it to our human routines, he concluded that it was important for good health to stick to a simple pattern: stay active during the day, and sleep at night.

These ideas were two hundred years ahead of their time. Even in 1962, principles which are now well established and widely known, such as the role of melatonin, and the disruptive effects of working night shifts or moving between time zones, were not yet understood. Ideas about the body clock were still regarded as niche, and the scientists who pursued them were viewed as working on the fringes rather than at the centre of the study of biology. But change was coming: a critical mass of research was beginning to form, and in 1960 a landmark gathering took place. The annual Cold Spring Harbor Symposium on Quantitative Biology adopted 'Biological Rhythms' as its theme, making it the first international conference dedicated to the subject. It included sessions on topics that were then at the cutting edge of research, such as circadian rhythms, animal navigation and homing behaviour.

It strikes Michel that his new idea might ride this wave of enthusiasm. He starts approaching scientific organisations seeking support for his new experiment. It's something genuinely new and important, he tells them. He knows he is not the first to turn to the cave for this kind of research; in the 1930s Nathaniel Kleitman carried out sleep experiments in Mammoth Cave in Kentucky, and his work is an inspiration for Michel. But this is different, he says. He will be alone for two whole months, and the effects on his body and mind will be systematically monitored from above ground. Unfortunately, most of the funders he approaches want nothing to do with it. Even his friends and colleagues are sceptical. He's too inexperienced, they say. It's too risky. Anyway, he's a geologist, and what has this got to do with geology? Some funding trickles in, but it's not enough. He will have to make economies. He decides to manage without basic items like a generator, waterproof boots and a fire extinguisher.

It's two o'clock on a summer afternoon when he makes his descent. It must have been an extraordinary leave-taking. The sky, the warmth of the sun, the breeze on his face, the sounds of birds: he turns his back on them all, knowing he will not experience them again for a long time, maybe ever. He knows he could die in the cave – there are so many ways it could happen. But he's in the grip of an obsession, heedless of all discouragement, rushing headlong towards the dark.

Walking in the rainforest the following night, I tilted my head back and further back to see stars in the canopy. 'Imagine what it is like when you are lost,' said our guide. She taught us a few basic survival techniques – how to know which way to go to find water, which kind of ants you could safely eat – but I knew it was a pretence. I would die, and the forest wouldn't care. There were brilliant points of blue and green luminescence in the trees and on the ground. I had never seen darkness so ardent with life. It seemed to contradict everything I thought I knew about night, about time, about so many of our human delusions.

I was woken in the early hours, when the four walls of the jungle rang in turn to a strange drumming noise: red howler monkeys beating the bounds with their great thundery voices. They came close to where we lay, then moved on, and after an hour or so the drumming faded away, but sleep was impossible now. I put my torch on its lowest setting and picked up my book.

The darkness of the cave is twofold. Like me, Michel has a headtorch, and he has installed a low-wattage electric lamp in his tent, so the ambient dark is tempered some of the time by artificial light, though he has only a limited stock of batteries and has to ration his use of it. But as soon as he moves away from his little camp, venturing out onto the moraine to fetch supplies, he is swallowed by the dark. He pauses and looks back at his tent, the lamp gleaming dimly within, and finds himself swept by a feeling he describes as love. When he retreats to the tent he is able to escape for a while from the sensation of infinite space imparted by the darkness around him. He is haunted by feelings of distance and alienation. 'The stars did not turn overhead,' he writes, 'and the days and nights did not succeed each other as they do over all the other tents in the world. My tent was outside the world. It was an entity in itself.'

Beyond Time was an unexpectedly good read. I kept telling myself I must switch off the torch and go to sleep, but then I would turn the page and find something funny or gripping and couldn't put it down. His account of that first expedition includes details of physical discomforts and hopeless domestic arrangements, an amalgam of the banal and the ghastly. Saving on the cost of new boots was a mistake: his feet get wet on the first day and stay wet for the entire two months. He is suffering from amoebic dysentery, picked up on a previous expedition, and his toilet is a bucket which he has to carry over the ice and empty into a hole. Despite the manufacturer's promises, the tent is not waterproof; it has also, inexplicably, been pitched the wrong way round. In his perpetual state of mild hypothermia he can't be bothered to pick up trash, so it piles up and partially blocks the entrance. He has brought saucepans and a supply of raw ingredients, but admits he has never actually cooked anything in his life; he thought it would be enough to bring a notebook in which a woman friend has jotted down some recipes for him – after all, how hard can it be?

Most mesmerising of all are the brief glimpses we get of his relationship with darkness itself. It affects his body in a multitude of ways which he observes and documents like a true researcher. His vision deteriorates, and his spatial awareness with it; he has bouts of dizziness and loses his balance when he goes out on the moraine. As the weeks go by he begins to lose his sense of colour, finding green and blue indistinguishable. He starts seeing hallucinations in the form of crowds of flashing lights; and when he shuts his eyes he is plagued by tinnitus.

Despite the appalling physical conditions, the real struggle is a psychological one, and he admits to occasional bouts of terror. The head torch skittering over the rocky walls is particularly alarming, as it was for me that day in Wales. In those moments 'everything seems to have a life of its own – the rocky walls, the great plaques of ice, the dancing shadows'.

In myth and legend, the cave – like the forest – is a primeval place, existing outside civilisation. A place of danger, where you must live by your wits. A place of solitude, whether you want it or not: a prison, a hideout, a lair, an anchorage. It's a place of discovery, where you might break through into a cavern glittering with treasure, or bright with paintings made by prehistoric people. Or you might encounter a beast, perhaps a supernatural one, or one thought to be extinct. It's a place of stories told round a fire; and you must keep that fire lit if you are to survive.

Caves are also places of ritual, altered perception and transcendent experience. In Ancient Greece, they were where the oracles lived, and across many cultures and times they have been sites of prophecy and shamanism. They are places where you can seek revelation, and enter states of consciousness that are not available in the ordinary world.

Research demonstrates that when external sensory stimuli are absent the brain fills the gap. This is one of the ways tinnitus can be understood: the auditory cortex is accustomed to a certain level of sound stimulation, and if there is a loss of hearing for any reason the brain generates phantom sounds instead. Something analogous happens with vision too. Some blind people see forms, colours and shapes even if they have no light perception at all. And for a sighted person, the experience of sensory deprivation alone in a dark cave creates a space in which the brain makes images of its own. These might be called visions, or hallucinations, depending on the context. They might be part of a complex of sensory experiences that contribute to a state of trance.

People who have spent days or weeks at a time on 'dark retreat', usually as part of a spiritual practice, frequently describe these visual experiences, which sound both fascinating and terrifying. On a physiological level, sustained darkness alters the functioning of the pineal gland deep in the brain, whose function is to produce melatonin, the hormone that governs the circadian rhythm. One theory is that the normal functioning of the pineal gland is disrupted and it begins to secrete DMT, the same hallucinogenic substance found in certain plants and used in the Amazon to brew ayahuasca. It's this psychoactive substance, the theory goes, that triggers the hallucinations or visions people have in the dark.

It's a controversial claim, and perhaps a reductive one. There are different ways of characterising these experiences – religious, mystical, psychedelic – but however we interpret them, they are waiting for us in the cave. Alone there for long enough, we will find our perception changed across all five senses. Spatial awareness will be transformed; the outside world will recede and lose its everyday hold, giving way to a dreamlike state which reveals other dimensions of reality.

Whatever the psychological stresses, he can survive them, declares Michel. He is an intelligent person with a trained scientific mind which can be occupied in study rather than left to run in other directions. It would be impossible, he thinks, for a mere *cobaye* or guinea pig to cope. But his confidence begins to crack as time goes on. The environment seems to him more challenging than the darkness faced by Arctic explorers, who at least have celestial phenomena to observe, whereas the cave dark is unchanging, a void in which he feels disembodied, composed entirely of his thoughts and fears. He spends a lot of time gazing at his reflection in a mirror, and feels his personality disintegrating. He senses that he is drifting close to insanity. The pages of the diary itself become a site of darkness; one of the later entries reads 'Writing with red ink seems to cheer me. I have had enough of black, quite enough.'

He shuts the book and lies down to sleep. He dreams that

his mother comes into the room and opens all the shutters, abruptly bringing the experiment to an end.

Soon after Michel Siffre descended into the eternal darkness of the cave, I was dragged out into the light after a long labour which left my mother weak and exhausted. Breech babies run in the family, but whereas both my daughter and my daughter's son were born by caesarean section, the protocol in 1962 was vaginal delivery if at all possible. I was born at home, in my parents' bedroom. At the midwife's request my father had climbed on a chair and removed the shade from the ceiling lamp, and it was by the light of this sixty watt bulb that she and my mother laboured together to bring me out alive.

The first dark we know is the dark of the womb, where each of us belongs until the moment of birth. It's generally assumed that we cannot remember this formative time, but it depends on what we mean by 'remember'. Understanding of prenatal development has changed radically in recent decades, and it is now recognised that the newborn baby is not a blank slate: long before birth it has sensory capabilities that enable awareness and learning, and the environment in utero shapes the individual in lasting ways. Our memories of this time are held not in the consciousness but in the body.

In my mother's womb I did not see the stars turning overhead. That space, too, was an entity in itself. Unlike the cave, however, it was perfectly designed for my survival and growth: I was held in warm amniotic fluid, nourished through the umbilical cord, sheltered from the outside world. My five senses developed within that protective space. My eyes opened for the first time in the safety of darkness.

A complicated birth is traumatic for the baby as well as the mother, and the sudden transition from dark to light is barely imaginable. Even the ordinary bulb in my parents' bedroom must have been blinding.

When Michel emerged from the cave at Scarasson, he wore goggles to protect his eyes. He was tearful and exhausted. One photograph shows him being carried away from the scene, cradled in the arms of a gendarme.

Time is the fabric out of which all life is cut. Without time, there is nothing. Yet it is the most slippery of concepts, notoriously difficult to describe or comprehend. It holds and structures all experience, but it remains subjective in spite of all the clocks we have built to measure it. There are days when we feel it racing ahead of us, and others when it seems to drag and dawdle.

The clocks that really matter are those inside our bodies, the ones that govern the essential functions we need to keep us alive: food, sleep, reproduction, the ability to locate ourselves in space. These clocks work not by signals beamed from satellites or transmitted from radio masts but by the cycle of light and dark, a repeating pattern that has been fundamental to life on earth from the very start. Without it, the clocks go awry. Hormones are released at the wrong times and our bodies don't know when to sleep or eat. We become anxious and depressed, our hearts get stressed, damaged cells go unrepaired, toxins are not cleared from the blood, spaces in the brain fill up with debris. We lose our alignment with astronomical time and become desynchronised. This is chronodisruption, and it leads to sleep and mood disturbance, obesity, heart disease, allergies, hormonal disorders and cancer.

Chronobiology must have been one of the most exciting areas of science to work in during the past fifty years: such rapid progress, such twists and turns, advances and reversals, such drama. In 1962 nothing at all was known about its workings before birth, for instance, but new research suggests that the molecular structures are already present and functional in embryonic tissue just a few weeks into gestation. Almost from the start, we have the makings of our own body clock, and there's a chain of mechanisms at work, synchronising and providing the developing fetus with circadian rhythm as it turns and floats in continuous darkness. The mother's principal clock, a tiny cluster of neurons situated in the hypothalamus, is entrained by the alternating pattern of day and night. In response she secretes melatonin and other hormones, which travel to the placenta. There they are translated into rhythmic signals that can be received and understood by the developing body clock of the fetus.

Equally astounding is the discovery that a fundamental timing mechanism is present not just in the hypothalamus but in every cell in the human body, and if a single cell is placed in a petri dish it will generate the same regular cycle. From conception until the moment of death, time is the stuff we are made of. We are living, breathing clocks.

Chapel

Being under a dark sky is like going back in time. The spaciousness, and the feeling of potential, remind me of early childhood, when for all my fierce fears and longings I was safely located, held at the centre of limitless space and time.

This was before I hit on the realisation that I would die one day. I was eating an apple at the time, and it struck me that just as I would come to the end of the apple I would come to the end of my life. I remember the moment, and the icy shock of it . . . but honestly, can it really have been an apple? More likely a Sherbet Fountain, or the magnificently named Everlasting Toffee Strip. I expect all the apples of legend and parable have got tangled up with the memory.

Either way, there was a before and an after. Before, I had no concept of my life as finite or conditional. I looked up at the night sky and thought of distance, but it was a comfortable distance because I was situated at its heart. I didn't know yet that the earth orbited the sun rather than the other way around; and the nineteenth-century hymns we sang in chapel were very slow catching on to Galileo. 'Jesus shall reign where'er the sun / Doth his successive journeys run,' went the opening lines of the old missionary number, which had a jaunty tune that enlivened a long spell in the pew. I sat snug there between my brothers and my parents, and the five of us sat snug at the centre of all that shivery distance and remoteness. The sun's successive journey had taken him round the other side of the globe for a while, but he'd be back tomorrow.

Meanwhile the darkness was also ours, in the woods and sheds and pillboxes where we played, and the wild places where my parents took us at weekends and in the holidays – the Peak District, rural north Wales – where we got to know truly clear night skies, and in particular the mercurial darkness that lay between a caravan and an outside toilet in the corner of a farmer's field.

One of the gifts of darkness was the enchantment of small lights that kept company with it: candlelight, torchlight, fairy light. In our house, candles were not the elegant creatures of the dinner table, but the stubby off-white kind that came from the hardware shop and were for emergency use only. However, there was another variety, treasured more than all the rest: the birthday candle.

At Sunday school, our birthdays were written down in a book and celebrated without fail. The focal point of the celebration was The Cake, which was made of cardboard, with holes punched in the top to hold the candles. One of the kind old ladies must once, in the dim past, have sat down at her kitchen table and fashioned it from a shoe box, and painted it white, and tied a bit of ribbon round it, and cut bits of sweet wrapper and stuck them on with Copydex.

At home our mum baked real birthday cakes, sandwiched together with strawberry jam, and decorated with hundreds and thousands or crumbled-up Cadbury's Flake. But strangely it's the cardboard cake that lives on most vividly in my memory. Halfway through a Sunday morning we would look up from our crayoning – we were always working on a frieze of the Good Samaritan or the Feeding of the Five

Thousand – and notice one of the old ladies, mysteriously busy behind the open door of the tall cupboard in the corner of the schoolroom, and a ripple of excitement would run through us, before the piano struck up for the familiar song:

> *Comes a birthday once again,*
> *Happy day, oh happy day,*
> *Through the sunshine, through the rain,*
> *God has brought us on our way.*

Then the smell of a struck match, and lo, there was The Cake, symbolic, eternal, a bit battered, and the candles in shades of pink, blue and yellow, colours that made your mouth water with the suggestion of sugar icing. Blow them out – *quick, quick* – so that they could be relit and the song sung again for the other kid whose birthday was the same week as yours. But there had been that moment, when the little flames had blazed especially for you, standing a bit wonky in the cardboard holes, quivering in the draught of our voices, and each wearing its brave halo of darkness.

ghastly white, looming out of the dark

You were alone in the house and it was getting dark. You noticed that the window reflected your own face, like a two-way mirror in an interrogation room. You imagined seeing something move behind you, and drew the curtain quickly.

Now you imagined someone creeping around outside. You went to the front door and turned the key.

As you climbed the stairs there was a fizz of presentiment at the back of your head. You forced yourself to turn, and there he was at the square of glass in the door. You knew him from the bus stop but he was different here. Ghastly white, looming out of the dark.

You had made this happen by imagining it. Now here he was, tapping the glass and saying *Are your parents in?* And here was you, giving the wrong answer.

Albedo

This is the hour when night says to the streets:
'I am coming'; and the light is so strange
The heart expects adventure in everything it meets;
Even the past to change.

— Frances Cornford, 'City Evening'

Music Room

Let us go then, you and I,
When the evening is spread out against the sky
Like a patient etherised upon a table.

Whoever is being addressed in the opening lines of 'The Love Song of J. Alfred Prufrock', it's a curiously irresistible invitation. I first read T. S. Eliot's poems at a formative age, and fell in love with their strange music and with the particular brand of stylish despair I heard threaded through them, which touched my own adolescent feelings in spite of all the great distances between him and me: time, class, education, wordliness. I guess that's why these lines always come to mind when I see a good sunset.

It has been said that this particular simile represents the very moment at which modernism gripped English poetry. The idea of the Very Moment is always beguiling, however simplistic. Never mind, it's a humdinger of a simile: gloriously dissonant, a dash of cold water in the face that wakes us with a start from the tired old quasi-Romantic dream of red and gold sunsets casting their shimmering paths over quiet seas. No, none of that works any more, because now some sickness has taken hold of the dream, and evening itself – that benign everyday thing – is a hospital patient, in mortal danger, drugged to sleep, waiting for the surgeon to take a knife to it.

But reading these lines today, over a hundred years after the Moment, I hear them suffused with their own brutal Romanticism, gritty and glittering. They grope through the mist of ether for the Sublime.

So I stand and look at a sunset, and these lines comes to mind, and I am enraptured all the same. Sunset is the very essence of the picturesque. How many images have been made of it over the centuries? Every one of them is a travesty. It's impossible to capture it truly in any medium, because it's fluid, it changes before our eyes. Sunset is not a state but a process, and each glance we take is unique and fleeting. The horizon blazes; it shimmers; it glows. Photographers adjust and readjust their camera settings; artists mix improbable shades of paint.

I am touched by a feeling of extravagance as I watch the colours spill and spread. They are the colours of nostalgia, evoking spring evenings on the streets and streambanks of my childhood, and after-dinner walks on beaches and hillsides, on holidays, in love. Memory speaks eloquently of the past in those colours. And they are also an invitation and a promise, presaging a gentle slide into the calm and enchantment of night.

Now the sun has set, and has sunk eighteen degrees below the horizon. This is the end of astronomical twilight, and the onset of true night. Colours continue to shift and vary, even if we are less aware of them. Sources of light swell and subside like tides, their different frequencies bathing the sky and the earth in shades of colour that can only be captured with special photographic techniques. The resulting images reveal that on a moonless night the sky takes on a reddish illumination, though this is lost on us as our eyes are unable to detect colour in darkness. A landscape under a full moon, meanwhile, can be revealed as remarkably similar to a sunlit one. It doesn't look that way to us, though moonlight can be surprisingly strong, even through a thin layer of cloud.

In our human frame of thinking, sun and moon are two separate sources of light, and the language we use to describe them is one of contrast. Each is gendered and lent a distinctive personality: the sun bright, hot and masculine; the moon pale, cold and feminine. Of course, in truth it's all sunlight, and the moon is a mirror – not just any mirror, but a priceless old mirror in the grand mansion of the night sky, a bit knocked about and with some of its silvering gone.

Some nights at the cottage I wake suddenly and think that one of us must have left the outside light on. In the first instance I argued against having one at all, but the nights get very dark here, and there are mossy uneven steps to negotiate on the way back from the log shed, and eventually I saw sense. But now I will have to get up and go downstairs and turn it off. On my way I tug the edge of the curtain aside, and oh, it's the moon! The garden, the lane, the trees, the sheep field beyond – all are visible, strange and pearly as if overlaid with a single sheet of tissue paper.

The earth has moved in its orbit and we lie in shadow, but far out of sight the sun still shines. The antique mirror of the moon reflects that sunshine and casts it down, muted but powerful enough to throw shadows and draw out textural detail. The gate across the lane is gradually coming off its hinges, and gazing out into the moonlight I can see the groove in the grass where we've forced it open day after day. The grass is silvery grey, and the groove a deeper grey. The lane is the grey of hammered steel, and I can see the silky grey leaves of ivy growing up through the old apple tree, and the complicated shadow it casts . . . and suddenly I notice a roe deer, standing stock-still in the lane, seeming to look directly back at me. Her night vision is not strong, but she has very acute hearing, and she is aware of me at the window. We stand locked together in an intense moment of attention before she turns and slips silently through the grey hedge.

It's not country lanes but city streets Prufrock walks at dusk: a place of foggy streets and sooty chimneys. Wherever they find themselves, poets tend to be haunters of the night, and the moon is one of their recurring subjects. 'The moon is the very image of silence,' writes the poet Mary Ruefle, and it's that quality of silence – made of distance, conditioned by darkness – that invests it with mystery, and makes it so readily available to metaphor. For the Romantic poets it was an irresistible subject, open to contrasting interpretations. Charlotte Smith, whose work is frequently situated in darkness, shadow or twilight, sees the moon as a place of refuge outside the span of human time, a kind of heaven reserved for those whose earthly lives are marked by misfortune:

> And oft I think–fair planet of the night,
> That in thy orb, the wretched may have rest:
> The sufferers of the earth perhaps may go,
> Released by Death–to thy benignant sphere;
> And the sad children of Despair and Woe
> Forget, in thee, their cup of sorrow here.

For Shelley, on the other hand, it is not eternal but mortal, subject to the same processes of decline and senility as we are:

> And like a dying lady, lean and pale,
> Who totters forth, wrapped in a gauzy veil,

Out of her chamber, led by the insane
And feeble wanderings of her fading brain,
The moon arose up in the murky East,
A white and shapeless mass –

Poets are often concerned with personifying the moon, but I'm more interested when they concentrate on observing it. Gerard Manley Hopkins was transfixed by it all his life. He was a habitual night walker, and wrote with great accuracy and feeling about moonlight, starlight and lamplight, as well as darkness itself. An entry in his diary describes a walk along the banks of the Wye at night, when he saw the river 'flush, swift, and oily, the moon streaking it with hairs'. On another occasion he observed moonlight 'hanging or dropping on treetops like blue cobweb'. In his writings we encounter a vertiginous spiritual darkness, a state of melancholy, loneliness and despair; but we also find a delight in the visual and sensory reality of darkness on which he turns his gaze. 'What you look hard at seems to look hard at you,' he writes, and indeed he does find himself regarded and comprehended by the dark.

Dorothy Wordsworth also knew and was known by it. She and her brother William were famously energetic walkers, setting out whenever they wanted, day or night, and apparently thinking nothing of it. Her journal entries are studded with moonlight and starlight, with the effects they create, and the way they alter and condition the darkness around them. In one entry, dating from their time living at Alfoxden in Somerset, she recalls: 'once while we were in the wood the moon burst through the invisible veil which enveloped her, the shadows of the oaks blackened, and their lines became more strongly marked'. Most of that walk, then, was moonless, and the rural paths of the Quantocks would have been very dark on such a night, but of course they knew them intimately. Having briefly observed with such clarity and precision this instant of revelation among the trees – the moon piercing the cloud, intensifying shadow and sharpening form and texture – she dismisses it as 'an uninteresting evening'. On another occasion, this time walking alone and apparently without fear, she is again struck by the drama of sudden moonlight: 'As I climbed Moss the moon came out from behind a mountain mass of black clouds – O the unutterable darkness of the sky and the earth below the moon! and the glorious brightness of the moon itself!'

If we only ever venture out by day, we miss out on these moments of celestial theatre, these transformative experiences where we see landscape and skyscape made new by the reflected light of that beautiful old mirror.

My father once got into trouble at school for claiming that Glenn Miller understood moonlight better than Beethoven did. A young man's perspective, perhaps; an inevitable preference for the smoky, sexy, two-step languor of the muted trumpet, and the sense that the summer night will go on forever, and that the opportunity it promises is limitless too. I've always thought 'Moonlight Serenade' comes to a rather incongruous finish – a flourish, a bit more volume suddenly, and that's it. It was the 1940s, and I guess this was the moment at the end of the dance when you had to say goodnight and go to your separate beds.

Glenn Miller's moonlight is shared and intimate; Beethoven's is huge and solitary. The lovely word 'albedo' describes the fraction of light reflected by a surface, and when I hear the Adagio – the first movement of 'Moonlight Sonata', the part everyone knows – it occurs to me that it is this reflected light Beethoven has in mind. I hear something in that series of arpeggios, and especially in the careful progression of chords that underpins them, that seems to equate to the act of observation or the taking of measurements. The whole effect is one of calculation, or reckoning, as if it were an attempt to wrangle an overwhelming sensory experience into a form that could be understood as mathematics.

There are experiences – apparently simple and commonplace, like the experience of moonlight that rapturises the level-headed Dorothy Wordsworth – that feel so intense they might destroy us. So much of art is an attempt to wrestle that feeling into a form, to translate or transmute it. Far from attempting to stir up feeling or exaggerate experience, the artist holds them in language, gives them a shape. Not to imitate reality, as Plato claimed, but to assuage it. We might think of it as albedo: not the ineffable thing itself, but a proportion of it reflected back at us. The poem, the painting, the piece of music, is the mirror.

It could almost be a defining image of Romanticism: Beethoven at the piano, struggling to find a way of writing about moonlight – a commonplace phenomenon, yet so potent tonight that it engulfs him. Then he sees it shining on water – the sea, perhaps, or a lake – and realises that the only way he can render this high-decibel experience onto the ear is to describe its reflection: the proportion of it returned by the surface of the water. The drama is modulated into a kind of musical equation.

It's nonsense, of course. 'Moonlight Sonata' is not Beethoven's title for his Piano Sonata no. 14 in C-sharp minor. At no time did he himself suggest any connection between the piece and moonlight; the association was made by others in the decades afterwards. It's a controversial nickname because of the way it seems to identify the sonata as a piece of programme music, and it has come to represent a fault-line between two musicological camps, one embracing the narrative potential of a composition and the other insisting on the primacy of the music itself.

The moonlight connection can be traced back to a writer called Ludwig Rellstab. In his story *Theodor: Eine musikalische Skizze*, three interlocutors discuss the work of Haydn, Mozart and Beethoven. At one point, Beethoven the Romantic is likened to the moon, in contrast with Mozart, the blazing sun of the Enlightenment. In this context a character, describing the Piano Sonata no. 14, says this: 'The lake reposes in the faint shimmer of the moon; the waves lap softly on the dark shore; gloomy wooded mountains rise up and cut off the holy region from the world; swans glide like spirits through the water with whispering rustles, and an Aeolian harp mysteriously sounds laments of yearning lonely love down from those ruins.'

Theodor hardly took the world by storm, and the picture it paints might never have taken hold of the public imagination if it hadn't been loudly disparaged by the musicologist Wilhelm Lenz. 'Rellstab compares this work to a boat, visiting, by moonlight, the remote parts of Lake Lucerne in Switzerland,' he complained, somewhat inaccurately. 'The soubriquet *Mondscheinsonate*, which twenty years ago made connoisseurs cry out in Germany, has no other origin.' But it was too late: audiences now heard a nocturnal mood in the piece, and the nickname stuck.

To my ear the moonlight comes and goes depending on the performance. Played badly it's a plodding and heavy-footed piece, but in the hands of a virtuoso it does become celestial. How I wish I had been in the room when Franz Liszt was asked to play it, and is said to have replied: 'Gladly . . . but put out the light completely, cover the fire, so that the darkness should be complete.'

The power of moonlight is the power of darkness, because although the moon is often faintly visible in the sky by day, it evokes no grand feeling, does not stir us into art, music or poetry.

The dark of a solar eclipse owes its drama to the light, because after all we are used to being temporarily out of range of the sun. It's the momentary intervention of dark into the brightness of day that is spectacular.

Our commonplace ideas of light and darkness as absolutes, existing in binary opposition, are eloquent misconceptions. They are not two separate states at all. There can be no light without dark, just as there is no music without silence, and no poem without white space.

In 2019 I went to see an exhibition by the artist Katie Paterson titled *A place that exists only in moonlight*. I took the train to Margate, a nostalgic trip because I had lived there for one mad year as a student back in the eighties. I was keen to see how much the town had changed, but my memories of it were mostly nocturnal, and as I walked along the seafront in the morning sunshine I could only really recall the times when I went to the beach at night, always in the grip of some or other turmoil, often talking to myself, once repeating Eliot's words 'On Margate Sands I can connect nothing / with nothing' over and over, as if they were a way of framing my own dilemma, though I couldn't remember now what the dilemma was, only the grittiness of the sand, the smell of seaweed and the wisps of fairground music carried on the breeze from Dreamland.

At the centre of the exhibition was a piece called *Earth-Moon-Earth*, based on the first movement of Beethoven's 'Moonlight Sonata'. I walked into the gallery where the music was playing. The room was dim, and the piano illuminated as if at a concert, though it was early in the morning and I was alone except for the invigilator on her folding stool by the door. I approached and stood as close as the cordon allowed. The piano stool was empty, there was no human player, but from where I stood I could watch the instrument doing its work, the keys depressed and released as if by invisible fingers.

Beethoven's adagio had first been translated into morse code, then transmitted by radio signal to the moon, which bounced it back to earth. When it returned, the code was translated back into a musical score. It was recognisable, but not intact. It had been disrupted and fragmented by its dark journey across space; gaps had opened up between chords, and elisions had modified the melody. This altered version was what I was hearing performed by the automated grand piano.

It was as if the moon itself were pressing the keys and filling the room with those carefully measured arpeggios, those reckoning chords. That dazzling celestial object, that priceless old mirror, had heard itself described in music and reflected it back to us. It knew nothing of Beethoven's intentions, or the arguments that swirled around the piece; it simply heard our human music and played it back, diligently but hesitantly. It was like a child in a piano lesson, fumbling the notes, making mistakes, doing its best. It faltered, broke off mid arpeggio, and there was a puzzled silence before it recovered and tried again.

For some long rapt minutes I stood there, and when I looked up an older man in a wool cap was standing by the door, and suddenly I thought of my dad, who had died recently, and the meaning of the piece shifted and translated again. Now I heard his own forgettings in those hesitations: the times near the end of his life when he would get confused, have to start a sentence again or trail off helplessly into silence. I thought about what it is to have feelings you struggle to articulate, and not to give up but to keep trying regardless. What we were hearing was not the piece itself, I thought, but the albedo. An ache began in my chest and spread upwards into my throat; I couldn't tell whether it was grief or consolation. In ones and twos more people came in from the foyer, drawn first by the music and then by the sight of the empty piano stool. We stood around in that moonlit space and looked at each other, and the feeling that passed between us then was awe.

Road

In my late teenage years, I was in the habit of missing the last bus home from the pub, which left me no alternative but to go the three miles on foot. The habit persisted in spite of the inconvenience it caused me, and looking back now I see I must have put some effort into maintaining it. Our own reasons are often illegible to us. Perhaps I was afraid of missing out on something during those last fifteen minutes in the pub, or perhaps I simply liked walking home in the dark.

On the first part of my journey there were streetlamps and traffic, but as I continued beyond the town limits it grew quiet; one by one the few cars turned off down smaller side roads or onto the driveways of houses, and I was really alone. At this point I would stop and take my shoes off. Ah, I spent so much of my adolescence in uncomfortable shoes, and for what? Anyway, my steps were quieter without them, and dangling them from one hand, swinging them nonchalantly as I went, made me feel more confident. On this long, straight section I would walk along the middle of the road, avoiding the shadowy gateways on one side, and the field hedges on the other. If a car did come, I would hear it and see it long before it posed any danger to me. Far more threatening were the unknown figures that might be hiding in those gaps and thickets.

School had been miserable; I had been through three

years of unhappiness there. (It wasn't till much later that I used the word 'bullying'.) I had felt isolated, and had learnt to retreat into the sanctuary of my own head, silent and intro-spective, gradually building an interior world that came to mean as much as the exterior one. I suppose I'm talking about imagination, but the word seems inadequate to describe that complicated private construction, the shape and texture and wholeness of it.

Now I was in the sixth form and suddenly everything was different. Life was one long riotous party, and I was invited. I didn't intend to miss a single moment. But I was so used to silence and solitude that I often found the social effort exhausting. I needed to retreat into that inner world too, and that meant spending time alone. The light of the pub and the dark of the road: now I recognise this as balance, but back then it just felt like chaos and contradiction.

I was reckless in those days – I went through a phase of accepting lifts from strangers, though one bad experience shook me awake from that foolishness. Friends would urge me not to walk, would offer me a sleeping bag on the sofa and assure me their parents wouldn't mind. Sometimes I took them up on it. But there was something seductive about the walking. I think it was the contrast, the potential to be two selves in one evening: wild and gregarious in the pub, then solitary afterwards, walking away from the lights and music with only myself for company.

I paused under the last streetlamp, leant on the post and took off my shoes. The drink made the buckles awkward. An owl called somewhere up ahead. I stepped off the kerb. The road was cool and steady underfoot.

Tent

Under cover of darkness, the unsayable can be said. When we are struggling to ask a risky question, confront a painful topic or articulate something private, darkness screens us from the gaze of others, hiding our vulnerability and making it possible to speak more openly than usual. Words are pushed out like paper boats, brave and fragile, carrying truths we would not dare to test in the full exposure of daylight.

A boy used to come to my tent after dark, night after night. He trod very softly and I never heard him approaching across the grass. He didn't carry a torch. He didn't follow the usual convention in that place of tapping on the canvas and saying 'Knock knock'. Instead I learnt to recognise the sound of his fingers finding the zip and carefully opening a space just large enough for him to crawl through. I would sit up in my sleeping bag, and he would fold himself to sit cross-legged in the narrow space beside me. Before either of us had said a word, he would light a cigarette, his face briefly radiant in the flare of the match. These were the only moments in which I saw him clearly, and the only moments in which our eyes met. The rest of the time, he was in shadow, and if the night was moonless the only source of illumination was the orange smoulder of his cigarette as he drew on it. Each draw was followed by a sigh, and these simple sounds, made from breath, served as preface and punctuation to an otherwise continuous narrative, or instalment: each visit an extension of the one before, each conversation picking up where the last left off.

I wonder whether it is accurate to call it a conversation, since he did almost all the talking. When he paused I would say something to encourage him to go on, but the dynamic was always the same: he was the speaker and I was the listener, and these occasional interjections of mine were designed to keep it that way, to maintain that status quo between us, and to support his lengthy and repeated attempts to get to the heart of his subject. It seems looking back that it was a monologue rather than a conversation, and that I was sitting on a hard chair in the wings, leaning forward slightly to ensure I heard every word, ready for the moments when his voice faltered and he needed a prompt. It looks from this angle, many years later, as if I played my part well, but then I remember a night when I drifted and lost focus and he suddenly said, 'Are you falling asleep on me?', and although I denied it we both knew it was true.

When all this began – and it went on for almost a month – I expected there to be a moment in which something would change. Perhaps he would lean towards me and we would kiss, or he would unfasten my sleeping bag and slide in beside me, or he would make one of these moves and I would rebuff him, because at that age I was very bad at knowing how I felt about someone until the moment of crisis. But after a while I recognised with mingled disappointment and relief that there would be no moment of crisis, and the tension fell away and it was safe to be together in the dark, close enough to touch, breathing his smoke and the scent of his damp hair.

He would talk and I would listen. He was circumlocutory to the point of evasion, and I can't remember anything he said, but I do know what it was about. He thought he was attracted to boys but he wasn't sure, and he wanted to be cool about it but couldn't because he didn't want to die. I tried to grasp the wasn't and couldn't – in my ignorance the whole thing seemed straightforward enough to me – but however much I wanted to be good at reading between the lines I knew I must be missing something. He went over and over the same ground – this was his one theme over the many hours we spent together – and it got harder to stay focused while he talked, and I started to feel irritated by his doubts and hesitations, and he must have noticed because he turned up less and less often and then not at all. I missed those nocturnal encounters, but not for long, and we drifted out of one another's orbit, but a few weeks later I was in the laundry room when someone was reading aloud from the newspaper about the American disease and how it had now reached Britain and men were getting sick and dying and they were calling it the *gay plague*, and I stood there with my arms full of wet clothes and thought of him and suddenly it all made sense.

Cave, 1988

'I no longer seek meaning in my life. I am, that's all. My existence has suspended its course. I no longer evolve, I no longer change . . . I melt into the eternal solidity of the environment around me.'

In 1988 a Parisian woman called Veronique le Guen returned to the surface after 111 days in a deep cave in the south of France. My own tiny subterranean epiphany, when I sat for a few minutes in the dark and experienced an altered sense of space and self, is nothing compared to the profound change she experienced in those months underground, alone and shut off from daylight.

Veronique was a keen speleologist and record-breaking cave diver who had made many adventurous and demanding expeditions. But this was different. It was an experiment, and she was the subject. She was required to record every change in her physical state, and to pass on information over a phone link with a research team above ground. There was a demanding regime of blood and urine tests, temperature checks, the weighing and measuring of food. These provided an element of structure in the otherwise drifting and amorphous weeks.

When I got to the end of *Beyond Time*, I wanted to know what happened next. I followed a trail online which led to another book, *Au Fond du Gouffre*, Veronique's account of her own solitary confinement underground. It was the latest in a series of experiments directed by Michel Siffre, who had long ago learnt how to describe his work in ways that brought in funding: these were survival experiments, he said, he was testing the limits of human endurance. His early research was of interest to the French military as they tried to work out how the sleep cycle of their nuclear submariners could be manipulated. It then caught the attention of NASA; this was a pivotal moment in the Space Race, and urgent questions were being asked about how astronauts might adapt to living outside the earth's cycles. NASA sponsored the longer and more sophisticated experiment in Texas in 1972, the six-month marathon Phil Cave told us about. Phil was by no means the only person to have a thing about Siffre: by then he had become a celebrity of sorts, giving interviews and posing for magazine shoots. In one photograph he reclines glamorously on his camp bed with electrodes attached to his face and chest, reading Plato.

Now he was keen to find a female subject, so that he could test the effects of prolonged darkness on the menstrual cycle. He met Veronique through her husband, also a speleologist. To her, this was an exciting opportunity. She had felt obscurely thwarted and directionless since childhood, and was happy to accept the mission. 'I hated my life,' she said afterwards. 'I felt I could do great things but I didn't know what they would be.' Here, then, was the great thing.

When Michel talked about his own cave experiences, and how they had altered the course of his life, she glimpsed the possibility that for her, too, this would be lifechanging. She was thirty-two, and was bored and unsatisfied in her secretarial job. She knew she was capable of more; she had a good scientific mind and there would be interesting and significant geological work to be done in the cave. And the scale of the challenge – the sheer length of time alone in that dark and silent place – would be a chance to test herself to the limit, which is one way of making something happen in a life that has stalled.

While Veronique le Guen was eighty metres below ground in the cave at Valat-Negre, I was giving birth to my first child and groping my way through the early weeks of motherhood. It was a long drive to the hospital, in the middle of the night, down the dark Devon lanes and then the brightly lit roads of town. Our car was a wreck, the brakes were not working properly, and we had to crawl towards red traffic lights to give them time to change. By the time we arrived it was an emergency – our daughter was footling breech, and the waters had broken – so I was put with haste into the solid dark of thiopental and she was born into the glare of the operating theatre. Then it was later, and I was wheeled on a trolley down corridors in search of her cry, my throat raw from the tubes and my head fizzing with stars from the drugs.

In childbirth, we experience wild changes in perception. In the overwhelming process of giving birth, and its immediate aftermath, the outside world recedes and we enter a dreamlike state which can be both disorientating and revelatory. From the moment the two of us met, we set up camp together somewhere outside time, all milk and pain and sleep and weeping and wild euphoria, no pattern or shape to it, while night and day unspooled like film off a reel into a tangled heap on the floor.

Time is shaped by our living. The things we do, the things that happen to us and around us. Extreme experiences – trauma, grief, falling in love – can bend and twist and warp it, as if it were melted and remade in the transformative heat of those experiences. It runs feverishly fast, or stretches like an endless hospital corridor. Or we lose track of it for a while, days and weeks becoming a meaningless blur.

But however intense the experience, and however compelling its effect on an individual's sense of time, the body is still governed by its internal clock, and by the cues it receives from the cycle of alternating light and dark as the earth moves around the sun. This natural inbuilt system regulates all our biological processes, telling us when to eat, when to reproduce, when to sleep and wake. By living in a cave, Veronique cut herself off from this pattern, and as a result her sleep cycle began to drift.

She was connected by telephone to colleagues above ground, and the protocol was that she would call them every time she woke, when she ate, and before she settled down to sleep. They were not permitted to call her, neither would they give any clue as to the actual time by their watches.

Extracts from the call log show how her sense of time deviated from the usual twenty-four-hour pattern. Her first few sleep-wake cycles were about the same as they would have been above ground, but then they began to detach and float free. Without the usual cues, her body sought its own rhythm. On one cycle she spent forty-three hours awake with a three-hour siesta, then slept for twenty-three hours. On another she was awake for fifty-nine hours, and asleep for thirty-three. She used the phone link to report each bedtime and each waking, along with the latest battery of test results, but the team above ground gave nothing away, and she had no way of telling whether or not she was still in synch. She experienced days and nights that were much longer than usual, and was incredulous when the call came through telling her the 111 days were up and it was time to return to the surface.

Theoretically, she could have used her menstrual cycle as

a kind of clock. It remained regular during the months she spent confined in the cave, so it would have given a good measure of time elapsing in that other world above ground. But she didn't trust this as a method of timekeeping. Her perception of time was so altered that she reckoned the experiment was affecting her periods and they were coming only eleven days apart.

Science has moved on fast since these early experiments. Michel Siffre's methods are controversial and his contribution to knowledge contested. He is generally credited with breaking new ground, but there is disagreement about how significant his work was, given the impossibility of generalising from individual case studies. It seems that his findings were compromised because he allowed his subjects to turn lights on and off, affecting the results. It is sometimes argued that he lacked scientific rigour, was driven by ego and enjoyed his Cave Man status too much. This led him to take risks: not only on his own account but with the physical and psychological safety of his other subjects.

By the time Veronique makes her descent, the conditions have improved. Her boots do not leak. There's a heater in her tent, a plentiful supply of batteries for the lamp and a freezer full of pre-prepared meals.

Despite these relative comforts, she finds the experience challenging. In her journal she writes not of terror but of rage. Her time is rigidly structured by a routine of blood and urine tests, temperature checks, the weighing and measuring of food. As the days pass, each very similar to the one before, it is increasingly difficult to tolerate their ceaseless and unchanging pattern, the invasiveness and monotony of it all. She starts to feel controlled and manipulated, a passive subject of someone else's experiment. Her resentment grows and grows until it explodes onto the page. 'I feel a wave of immense aggressiveness that dominates my spirits,' she writes. 'I look at each of the instruments of my torture: equipment to take samples, analyze, count up, manipulate, pierce. A crazy desire overcomes me to smash and destroy everything.' Until now she has been on good terms with Michel, but she starts to feel he is enjoying exercising power over her, torturing her from above ground. She makes a series of furious sketches. One is a self-portrait, showing her strapped and bound with wires and probes. In another there is a shelf of books on sad-omasochism, with his name on each of the spines. A third depicts a dartboard in the form of his face, wounded by many darts.

After the experiment was over, she said repeatedly how glad she was to have been through it, and even spoke of her longing to be back underground. She paid tribute to Michel, calling him 'the great *ordinateur* of the experiment, the one to whom I owe this sublime and trying period of my life'. She published her journal, adding an epilogue in which she looked back on it as a beautiful experience, before reflecting 'but I must admit that I sometimes experience periods of complete psychological disconnect where I no longer know what my values are, the purpose of my life, etc.'.

Of course, no personality cult grew up around her. There were no photographs of her reading Greek philosophy. The press described her as 'an attractive young secretary', and reminded readers that she was the daughter of suburban shopkeepers. They mostly neglected to mention her previous achievements. The intense media attention soon fell away. And it must have been hard to pick up the threads of her life – the life she said she hated – having invested so much in that momentous descent into the dark. At times the rage flared up again: 'The genius Siffre had a brilliant idea,' she said, 'to lock a woman in a cave. Even the most hardened criminal would never have thought of it!' Those who knew her saw her struggling to assimilate the experience, and seventeen months after emerging from the cave she took a large dose of barbiturates and ended her life.

Who knows what drove her to such an extreme conclusion? Such questions never have easy answers. Her husband said that she had not been properly prepared for the likely psychological effects of the experiment, and that her time in the cave had left her with 'an emptiness inside her which she was unable to communicate'. Michel, on the other hand, denied that there was any connection. Despite having documented his own struggles to maintain his mental health while underground – feeling his personality splitting apart, considering suicide – he insisted that the reason for her death was purely personal. He pointed out that she and all his subjects were willing volunteers. 'They choose themselves!' he declared. 'And I transform their will to be known, to become glorious in scientific experiments. It's why we have always had a success. We have never had a failure in any of my experiments.'

The subterranean explorer is traditionally depicted as a man. The helmeted figure descending a narrow shaft into the unknown is male by default; adventure, risk and discovery are automatically his. These assumptions may be more recent than we think, however. There is some evidence that in prehistoric times it was principally women who used and controlled such spaces, and a study of handprints found on the walls of Palaeolithic caves in France and Spain found that most of them were made by women's hands.

In 2023, when Beatriz Flamini emerged after a record-breaking 500 days alone in a cave in southern Spain, she surprised reporters by calmly describing it as an enjoyable and satisfying experience. They must surely have been hoping for something more dramatic. But Beatriz was in charge of this expedition: the ambition was hers, and she was the *ordinateur* who planned and designed the whole thing. When asked how she had managed to keep her equilibrium during the long months underground, she simply replied, 'I was where I wanted to be and so I dedicated myself to it.' Perhaps the key to surviving an extreme experience like this really is a sense of ownership. Michel himself said, all those years earlier, that no *cobaye* could be expected to do it.

Veronique's situation was very different. She was excited by the prospect of contributing to science, but in the end it was someone else's science, and her role was to do as she was told: insert a probe into her rectum, spit into a test-tube, draw blood for testing, attach electrodes to her face and scalp, over and over again for III days. The secrets of her body were drawn out into the light to be scrutinised by others. When I think of her in that cave, I cannot imagine being more alone yet less private.

I find myself haunted by her, that dynamic and adventurous woman, who embarked with such enthusiasm on her long expedition underground. Claustrophobia alone means that I cannot imagine doing what she did. My own love of darkness is qualified and provisional in ways that hers was not; she knew it intimately, and went into it without the terror I would certainly have felt in her place. But in the cave the darkness was combined with profound solitude. She was cut off not only from other human beings but from all living things. She could not, as I can this evening, step out into a cool night and raise her eyes to the sky to see the stars and the spaces between them. It was a form of exile: the most extreme form imaginable, because the distant homeland she dreamed of was not a single country, or even a continent, but the universe.

No wonder she was jubilant when she was told the experiment was over. But even as she thought excitedly about her return to the surface, she recognised that she was losing something too. She lit candles on the stalagmites she had come to know so well, and addressed them for the last time. For many weeks she had been alone in this abyss where time had lost its meaning, and tomorrow she would return to the surface and resume her life, while the stalagmites continued to exist on a different timescale altogether. How hard it must have been to reconcile one reality with the other. 'I praised their bearing, their elegance,' she wrote. 'I didn't mention their age, because like Dorian Gray they are not accessible to time. I thought, not without pangs of the heart, in one of those lucid but fugitive visions of life, that I would return with white hair, my face wrinkled like an old apple: I would find them again, these staunch companions of youth, just as bright and fine as they were today.'

Darkness surrounds our planet, but it also perforates and penetrates it. The ground beneath our feet is not solid but riddled with caves, tunnels, mineshafts, sinkholes – some naturally occurring, others human-made. We too are penetrated by it, and our fear of darkness is also a fear of the hidden inner parts of our own bodies, where essential but mysterious processes are taking place and things can go wrong at any time without our knowledge. Our dark interstitial spaces can be mapped, and occasionally we can look at them in images generated by X-ray, ultrasound or MRI. This form of self-knowledge is very new. But without the expertise to interpret those images, they remain abstract. What we see is something indistinct, almost schematic, not easily reconciled with our own lived experience of our bodies. It's like looking into a dark room where objects reveal themselves only as forms, or staring across a field at dusk at a shape between trees, and hesitating: should I be afraid of this, or fascinated by it? Then the doctor says, 'This area, here – this is where the problem is.' And we nod, though we don't really understand what they're pointing at. Darkness is not an external phenomenon, something out there, to be feared or admired from a distance. It's part of us. We are part of it.

Since prehistory, there has been an association between the cave and the womb. Many cultures regard the earth as mother and source of all life, and situate in her dark interior spaces the vital mysteries of sex and procreation. The Utroba or Womb Cave in Bulgaria is thought to have been carved about three thousand years ago into the shape of a uterus, and to have been an important shrine to the Thracian people who lived here at the time. The cave is entered through an opening resembling a human vulva, and inside is a stone 'altar' shaped like a cervix, positioned beneath a hole in the ceiling. On certain days of the year the shaft of light from above appears to lengthen and then penetrate the cervix-stone in a phenomenon that may have represented fertilisation.

Until the development of imaging technologies made it possible to see into the dark interior of the womb, it had to be visualised with the help of illustrations. In her book *Birth Figures*, Rebecca Whiteley considers a series of these early anatomical figures – drawings, paintings and woodcuts – which were used to help midwives and doctors understand different presentations during pregnancy and the methods they would need to employ in order to achieve a complex birth. A breech or transverse presentation was extremely dangerous for both mother and baby; feet-first or footling breech was one of the most dreaded presentations, because of the high risk of cord prolapse, and of the baby becoming stuck in an undeliverable position. These anatomical figures, and the insights they afforded, must frequently have made the difference between life and death.

When I went to antenatal classes in the 1980s, the teacher used a newborn-size doll, complete with detachable umbilical cord and placenta, and a model pelvis made of hinged plastic. (She had an absent-minded habit of articulating the pelvis in her hands as she talked, and with each movement back and forth it gave a click that so got on my nerves that I thought I would have to heave myself off my beanbag somehow and leave the room.) The figures in Whiteley's book served a similar educational purpose, though they were designed for those assisting in the delivery rather than those giving birth themselves. They are mesmerising to look at. The womb is imagined as an inverted flask, or as a garden, or as a world, and in these spacious environments float the unborn, plump and merry, often in acrobatic poses, with mature faces and knowing expressions.

Until the last century, childbirth was the leading cause of death for women between the ages of fifteen and forty-five, and in many parts of the world it is still appallingly dangerous. How grateful I am for every mitigation against the danger. For the illuminating power of imaging technology. For the doctors and midwives who learnt to see with their hands, and the medical illustrators who helped them. For the plastic pelvis and the demonstration doll. For the bright lights of the operating theatre, and the dark of the anaesthetic.

I stare at a page in *Birth Figures*. Here's one that looks more like a toddler than a fetus, and at first glance appears to be wearing a powdered wig. He is shown in the single footling breech position I recognise from my own ultrasound scan back in 1988: one leg extended downwards into the neck of the flask and the other bent, as if caught mid stride as he runs carefree across a lawn in pursuit of a butterfly. He is smiling, as if the whole business of getting born is going to be a breeze.

light burned your scalp

You can't say what triggered it. You were at a party (there were so many) but you'd only just arrived so you can't have had much to drink. You just had this feeling that you were not real. It was like suddenly waking from a dream. In the dream you were part of the real world, but then you woke and found you were not.

The waking was brutal. Light burned your scalp. Time buckled and warped, things happened a split second before they happened. Your hands felt big and clumsy, and when you touched your face it seemed to be made of something inert, like papier-mâché. You were losing touch with your body, becoming detached and incorporeal.

You could hear yourself screaming but it wasn't something you were doing. People were running around, someone turned the music off, you were caught by your flailing arms and held. Next thing you were lying down and begging to be left alone in the dark. The light was switched off and the door closed.

You began to feel better. You could hear the party getting going again downstairs, and it felt good to be here rather than there. In the dark, putting yourself together again, piece by piece.

What was it all about? Looking back on that time you see you were trying to split yourself in two. Things had happened that must never happen again. They must never even be thought of again. The answer was to strike one half of yourself from the record.

The next day you had to pretend to be ill. It provided an answer for your puzzled friends and their concerned questions. Must have been going down with something. Sorry. Sorry. You didn't tell them it had happened before. You didn't have the words. All you knew was that the cure was a dark room where you could fold back into yourself and be real again.

Tenebrism

– O remember
In your narrowing dark hours
That more things move
Than blood in the heart.

— Louise Bogan, 'Night'

Hospital

When it became clear that my mother was dying, an orderly came and wheeled her bed into a side room. I had been at the hospital every day since her stroke, but I had not noticed this antechamber, where light was dimmed and sound muted. Patients were taken to spend their final days or hours here, free now from tubes, needles, tests and medications, sequestered away from the noise and bustle of the busy ward, the clamour and undersong of the monitors and breathing machines, the bang of steel drawers on the nurses' station, the awkward chatter of visiting relatives.

I couldn't help noticing as I locked my car and walked towards the doors of the hospital that the weekly parking ticket I had bought from the machine would run out the following day. If she died tomorrow it would in a sense be my fault. Such meanness in me to buy a week rather than a month, or even an annual season ticket. I should have brought a suitcase full of coins and stood feeding that machine till it could take no more. But it was no good, she was not going to recover, and I didn't wish her to be like this forever. So I propped my weekly ticket on the dashboard, locked the car, and walked on through the automatic doors and down the dazzling corridors and along the row of parked beds and through the blue curtain into the dim room beyond.

It was very like the screened and secluded space where

I was taken twenty years earlier to recover with my newborn son in my arms. Even the curtain looked the same. It could almost be the same kind nurse in the plastic apron who brought me tea and said, 'It's lovely and peaceful in here.'

I knew that in childbirth darkness induces the body to produce more oxytocin, making contractions more effective and easing the baby along the birth canal. Might it also help in the process of leaving the world? And how could I ever bear to leave this twilit room without her? The lights were so low I had to bend close to look at her face, and I saw then that she had already stopped labouring and slipped into a state of calm. The fever had passed, her breath was slow and shallow, and she was pausing here to rest before moving on.

Loft

Clearing our parents' house took repeated visits, and recurring encounters with the same cast of objects. It was hard to think of a method for such an intimidating task, and without discussing it we began to empty the cupboards and put things into groups. Certain places in the house became the sites for these assemblages. There was one on the dining table, another in the garage, a third on the kitchen floor (very inconvenient, since the kitchen was so narrow). Each had its own distinctive character – things made of glass, textiles, electrical items and so on – but many objects are composite, and definitional boundaries are notoriously open to interpretation, so the organising principle frequently broke down. Then there were all the things that couldn't be made to fit anywhere. I dithered helplessly with a wooden box stacked neatly with shoe brushes and tins of Cherry Blossom polish in several shades, taking it from one site to another and back again, before acknowledging that it belonged in none of them, and that despite the fact that my father had kept it in weekly use since the 1950s it now qualified only as rubbish.

Recently Dad had been making his own assemblages, according to organising principles that were mysterious to the rest of us. In his mother-in-law's old china cabinet, displayed alongside a rose-patterned bowl and a jug commemorating the marriage of Victoria and Albert, was an egg-slicer.

I don't think he could remember what this peculiar object was. Perhaps it was intended to be decorative, he wasn't sure. I imagine him standing doubtfully in the living room, dithering like me, before deciding to classify it as an ornament.

When I handled it, the egg slicer was intoxicatingly familiar. It was made of plastic that had once been white but had gone a bit yellow with age, and it had a little handle you turned to bring the row of metal blades down on the hard-boiled egg. The only time I remembered it being used was when we were making packed lunches for the annual Sunday school outing. I would be in the kitchen with the old ladies, doing bread and marg for the sandwiches, spreading some with fishpaste, some with Primula cheese squirted from a tube, and some with sliced egg and salad cream. There would be an air of wild anticipation that morning, as the greaseproof bags were stacked in a box to be carried to the coach. Dad was out at work, of course, but perhaps some instinct told him, all these years later, that this puzzling object had played a part in a key ritual of our childhoods and was therefore worthy of a place in the china cabinet.

The most daunting part of the house was the loft, which I knew was full of boxes. My parents did not like to waste things, and when they replaced something – a toaster, a vacuum cleaner – they were in the habit of keeping the broken one just in case; we had already put three old radios onto the heap in the garage. They had also carried with them, from house to house during their long marriage, a trunk containing all the congratulations cards from their wedding, along with the invitations and the thank-you letters and the receipts for the catering and the spare paper napkins. In there too were stacks of carefully folded artefacts from our childhoods – drawings, stories, school reports, certificates for swimming

and music exams and cycling proficiency – and all the corre-
spondence relating to the deaths and funerals of their own
parents. I had seen some of these things on occasion, and had
glimpsed deeper layers: small diaries or notebooks, and dusty
envelopes, some addressed in unfamiliar handwriting, others
tied with a scrap of ribbon. If there was anything I wanted
in the loft, it would be found at the bottom of the trunk –
some long-forgotten item that would solve a family mystery
or reveal some aspect of my parents' lives before my brothers
and I were born.

I climbed the creaking aluminium ladder and poked my
head through the hatch, feeling around for the switch. It
didn't work. Something had gone wrong with the electrics.
The space was packed with warm dark air and a familiar
fibrous, mousy smell. I hesitated, with my head in the quiet
of the past and the rest of me in the chaotic present. I would
have to do this by torchlight.

The first boxes I investigated – big cartons that had once
contained a television, a double duvet, a cooker hood – were
empty. I squashed them roughly and pushed them down
the ladder onto the landing. So far so good. On to the Pick-
fords packing cases, labelled with permanent marker: SPARE
BLANKETS. PICTURE FRAMES. CHRISTMAS DECORATIONS. All
empty. I flashed the torch into distant corners of the loft.
It looked surprisingly bare. Where were the battered old
folding chairs, the wallpaper pasting table, the tins of Lego
and marbles for visiting grandchildren? I seized at boxes that
were so light I knew there could be nothing in them, threw
them down and moved on feverishly. Nothing. Nothing. Old
bubble wrap. Nothing. My heart was in my mouth as I staggered towards the trunk, my back bent under the slope of
the eaves, trying to step on the joists and avoid the gaps lined

with fibreglass. I knelt and fiddled with the two metal hasps, held my breath as I heaved the lid, but I already knew before I shone the torch: there was nothing inside.

It was hard descending the ladder because of the sea of empty boxes on the landing. Dad had got rid of the lot. He must have made so many trips up and down, with his arms full of stuff, still fit and nimble in his late eighties. He never mentioned it, but he was on a mission. I would never get to untie that ribbon and read those old letters. And that was only right, really – they weren't intended for me. My eyes were wet but it was probably the dust. The trunk was too heavy to bring down, I'd leave that problem to the new people.

Darkness feels different to the newly bereaved, when we enter into a changed relationship with time. When my dad died I lay awake at night, going over and over the events leading up to his death, then plunging further back to think about my mum's death seven years earlier. His death had a double impact: it finalised hers. While he was alive her presence could be felt in the rooms they had shared and the objects that furnished them: a coffee mug she liked to use, a recipe book with her handwritten notes on some of the pages. For a time her piano still stood there, with its lid shut like the lid of a coffin, before Dad gave it away. But now he was gone too, and the recipe book and coffee mug, and the egg slicer and the shoe brushes. Awake in the dark I was pitched back further still, into childhood, when the two of them were part of a fixed and eternal pattern, like a constellation. Even in my fifties it was hard to assimilate the loss of them, not only because it made me sad but also because it didn't seem to make any logical sense. There had been other deaths, of course: illness and accident had taken its toll of friends and family over the years. But this was different somehow. If they could die, then I will die, I thought, and felt a shift inside me like a gearwheel clicking forward over another cog, as this fact – which of course I knew, had known since I was very young – tightened its grip. It seemed to be taking me a lifetime to really believe it.

In the cupboard under the stairs, among the tins of scouring powder and Brasso, I found a misshapen candle stub on a brown saucer. In ancient times, candles were used as clocks: meticulously divided into equal sections, and shielded from draughts so that they burned at a steady rate. But as I have lived through the middle of my life and beyond, I have come to see *every* candle as a clock. These days, when I blow one out, I notice how much time has melted away and guess how much more there is left. Even new candles, stacked pale and smooth in the drawer, can be counted and used as a measure of time passing. It occurred to me that this stub – deformed by its own liquefying into the shape of a dowager's hump – was eloquent on the subject of mortality in a way that a lightbulb could never be.

I decided to take a break from the house clearing by revisiting some of my childhood places. I wasn't expecting to be back in the area much in future, and felt I should take the opportunity now. I walked past the house where I was born, and the almost identical one across the road we moved to when I was a year old. Both had been extended and altered and it was difficult to reconcile them with the pictures in my head. The roadscape, though, was just as it had always been – the corners and cul-de-sacs, even the drain cover where I lost a sixpence in 1971. (I crouched and peered in, just on the off-chance.) But I had the feeling I'd been back often enough and would probably not do so again. Then I went into town and wandered aimlessly around the streets, gazed at the shiny metal towers of the breweries, found the old cinema reopened as a cat café, got lost in the shopping precinct.

What I needed, I said to myself, was a proper getaway, ideally to one of the places I loved best as a child. I craved the sweetness of nostalgia. So I locked up the house and left it to its introspections, and drove north to Matlock Bath, crates of china for the charity shop rattling on the back seat.

Matlock Bath was the closest thing we landlocked Midlands kids had to a seaside resort. There were amusement arcades, and fish and chips, and a shop selling candyfloss and sticks of rock, and coloured fairy lights strung between lamp posts on both sides of the river, their reflections dancing excitedly on the water. We did have a set of fairy lights at home, but they only came out for two weeks at Christmas and were then carefully packed away again in tissue paper and put in the loft. These riverside lights stayed on all year. Such extravagance! I felt their allure again today as I drove into town with my carload of grief and sophistication. It was only four in the afternoon, but dark comes early in this narrow ravine.

I had booked a room on the Parade, and after the inevitable fish and chips I went for a walk along the river. There was a long row of motorbikes parked up outside, but the street was quiet. It was raining steadily, and the air sang with the sound of water meeting water. I hadn't realised autumn had come until I stood on the ornamental bridge and saw the empty benches lacquered with rain under the lamps on Lovers' Walk.

In the old days, there were fireworks during wakes week, and boats hung with Chinese lanterns. Nowadays there's a festival each autumn, and enthusiasts work all year building elaborate designs and fitting them with lights, to be mounted on boats and rowed up and down the river. At weekends the town is thronged with visitors, and you have to buy a ticket to stand on the bank and watch. But tonight it was Tuesday and I met no one as I walked the riverside back to the hotel. I passed the puddled ground of the fair, where the empty rides and illuminated signs augmented the dark. Leaves were falling, and rain was pasting them like notes to the pavements and the motorbikes, and I thought of our assemblages back there in the locked house and the notes we had stuck on them, and although I was glad not to be in those forsaken rooms tonight I wished I had at least closed the curtains to make it feel more like home.

Lead Mine

When Simon our guide throws the big switch on the wall, I feel the sudden darkness like a quenching draught of cold water, drunk down good and fast on a hot day.

He strikes a match. It rasps twice, three times against the box before igniting, and hesitates at the wick before it catches. The small light of the single candle clears a space for itself, as a drop of vinegar does on a saucer of oil. It opens a circle in which we can see each other's faces, and Simon's hand around the candle, the thumbnail chipped and blackened, and the gleam of his hard hat, and our boots on the stony floor.

For someone who dislikes being underground, I find myself returning surprisingly often. I can't explain it, except that when I woke up this morning it was still raining, and I remembered the old lead mines. Once you leave the Parade and head out of town, the hills are riddled with them. This picturesque part of the Peak District may not at first glance look like a post-industrial landscape, but the legacy is there in the form of hundreds of mineshafts, some glimpsed under iron grids, others concealed in dense woodland. It's there on the ground, under your feet and around you as you walk, in the piles of rocks that turn out to be spoil from the mines, and the unusual wild plants like spring sandwort and alpine pennycress that thrive on the high lead levels in the soil.

I have heard that in some parts of the world early miners lit the shafts with jars of fireflies. Can this really be true, I wonder? Even with the candle I can't see much outside that magic circle, but when I step closer to the wall of the tunnel I can just make out folds and angles in the rock, picked out in light and shadow, and a mysterious cluster of pale forms like tiny stalactites. 'Snottite,' says Simon. 'It's a biofilm – grows everywhere in here.'

He has lit the candle to demonstrate how this tunnel would have looked to the men working the lead mine in the days before electricity. It isn't a true demonstration, since this is a modern paraffin candle and the ones the miners used were made of tallow; the light they gave was more dingy and indistinct. They were kept in a tin box, to keep them safe from rats, and lit not with a box of matches but with flint and cotton rag. Miners had to buy their own candles out of their wages, and they were coloured green to indicate that they had been bought at a special subsidised rate and were only to be used in the mine, though they were often smuggled home and used illicitly to light the women's work in the long winter evenings. Each one burned for two hours; if you kept count of how many you used you could time your shift by them. The flame was easily blown out in a draught, and if you were careless or inexperienced enough to drop the tin box on the wet floor in the dark you would have to kneel and scrabble around until you found it.

Like the rats' tails that once lit our cottage, the miners' candles were made from mutton fat, melted down and poured into a metal tube, with a piece of rush pulled through it to act as a wick. They smoked and guttered and stank of sheep. I suppose you got used to the smell, I suggest. More than that, says Simon, you came to love it. Darkness was your element in here, and you learnt to respect it; but any small light you could conjure to cut a space in it was precious and hard-won. You would scratch your initials on the tin box and treasure it.

This brief foray underground is not very challenging, even for someone of my disposition. We're sticking close to the surface, making our way from one spacious cavern to the next. But then we follow a tunnel into a narrow passageway where the roof is low and craggy, and I feel a familiar clutching sensation behind my ribcage, spreading up into my throat: claustrophobia. My body does not want to go any further. Simon switches on his torch and waves it around to show me there's plenty of space, actually, but the torchlight glancing off wet rock makes it worse. I bring my nerves under control by imagining I'm a miner, with my green candle stuck to my leather cap with a blob of clay, leaving both my hands free for the work. It takes skill to move through these limited spaces without dislodging it, and if it does get knocked off my cap I'll have to find flint and rag and relight it before I can continue. With time and experience, I tell myself, I will develop a mental map of the space I'm working in, and adapt my proprioception – the brain and body are very good at this – and before long it will become second nature to navigate safely and allow enough space above my head for the flame to pass unscathed. Even as I have this thought, I crack my elbow on a spur of rock.

Meanwhile, Simon is telling me about the way these lead mines were constructed: the sinking of shafts, the digging of tunnels, the driving of levels. There's one kind grimly known as a 'coffin level' that was built wider at the top than the bottom, and just big enough for a man to work in. Then there were soughs: passages cut specially to drain groundwater from the mine and carry it safely away to the surface. The labour it took to open up these places, and then to work them, is credited with producing a certain temperament among the locals: stoical, self-reliant, phlegmatic.

No doubt the old miners would think me mad, paying good money to shuffle along this damp passageway, but lead-mine tourism has a surprisingly long history. In one account a visitor to the area in 1687 spotted what he thought were wells but was told they were mine shafts. He asked to be let down so that he could take a look. No sooner was he lowered into the mine than his two candles blew out. The miner acting as his guide said casually, 'It is nothing but a damp,' and went back to the surface for more lights, leaving him alone in the dark. Feeling short of breath he crawled back to the shaft, gasping for air, until the guide returned and assured him 'such damps were not killing'. It sounds to me like the holiday from hell, but he insisted it was all worth it in the end: 'I went back to the mine, where their constant lamps and candles, which they lighted for my sake, did make the glittering of the ore very pleasant to me.'

What were the forces that drew people in from the hills and fields to these dark rifts and fissures in the limestone? Economic necessity, certainly. Beyond that, the dream of real wealth, enabling a life of luxury, and the chance of social mobility. When the painter Joseph Wright of Derby became engaged to the daughter of a lead miner, for instance, his family objected on the grounds that she was in 'an inferior station in life', but in spite of the class difference her father seems to have made a lot of money from his venture, and the marriage went ahead.

These caves have been worked as mines since Roman times, regardless of the graft it took to open them up, remove the minerals and bring them to the surface. During the industrial revolution, investors and speculators came here from all over Europe, and the competition was feverish. In the confined space of the mine, where the work was gruelling and the stakes excruciatingly high, scuffles broke out, and in the commotion all the candles blew out too, and the fight continued in darkness until someone found and struck a flint. On a larger scale, arguments between rival mines could escalate dangerously. In one case a dispute over ownership of a profitable vein ended in deadly violence, with each side setting fires underground in an attempt to smoke the other out.

Even without these conflicts, a lead mine was an inhospitable place to work: cold, wet, dangerous and very dark. Yet adventurous young men went eagerly into them, not only to earn a living but also to chance their arm. It was cheap to start out, and a labourer working in one of the larger mines, or on the railway or canal, could have his own small mine on the side. It might not look much, but there was just the possibility that he would strike it rich. Lead wasn't the only profitable find; you could do very well for yourself out of minerals like pyrites, ochre and fluorspar. There was always a story doing the rounds in which someone had broken through into a vein of ore and made his fortune, and these stories sustained the small mine owner through the months of backbreaking and thankless work underground. It was a grand gamble, made all the more unpredictable by the darkness: the candle burning on your cap might suddenly reveal the life-changing glint of ore. On the other hand, you might well spend weeks driving a crosscut into a vein, only to go in with your light and find it full of nothing but the same old rocks.

Disappointment was not the only risk for the prospector. The names of old mines – Hazard, Goodluck, Never

Fear – reflect an awareness of the dangers. There were so many ways in which you could be killed or maimed: rockfall, explosion, drowning, tumbling down a shaft. In the narrow and poorly ventilated workings there was always the risk of choke-damp, and the miners used to test for it by holding out a long piece of wood with a bunch of candles tied to the end. The industrial historian Nellie Kirkham recalled this method being used in Derbyshire as late as the 1930s, when men were clearing out an old sough.

I already knew about the soughs, from my conversations with my old friend Doug Nash. I met Doug near the end of his life, when I went to see him about a gravestone I wanted to write about, in the churchyard close to where I grew up. A drowned boy was commemorated there, and I knew that Doug had been present at the drowning, many decades ago – indeed, he had himself come close to death. I wrote to him to ask him whether he would be willing to tell me what had really happened, and he invited me to visit. He was frail and forgetful, but full of wonderful stories, most of them situated in the flooded underground spaces beneath the field and hills and villages of the Peak District. He claimed to be unaffected by that childhood experience, one that I knew must have been traumatic, and yet he had chosen to spend his working life diving and wading through subterranean water, risking everything, in and out of his depth.

There are miles and miles of these soughs, he told me, largely unseen and unknown except by the small community of enthusiasts who explore them – cavers, industrial historians, urban explorers – and bring back pictures which show the breathtaking workmanship of the soughers: vaults, arches, exquisite brickwork. In places they resemble subterranean canals in scale and ambition. On one there's even a set of lock gates; I remember Doug's trembling finger pointing at an old black-and-white photograph of them, with the initials O. M. inked onto the frame: Operation Mole, the name of the venture he set up with a couple of friends, commissioned by landowners and planners and utility companies to survey and map the Derbyshire underlands. That world was everything to him; he worked underground all week, and then went down at weekends too, for fun, exploring caves and mines with fellow enthusiasts. On several occasions he came close to drowning, as he had before. But it didn't put him off; he loved those places.

Doug and his mates had powerful head torches to guide them through the soughs and levels, but all the work of digging the soughs was done by candlelight. He told me how the soughers would prepare to open a channel in the rock by hacking off the outer layer, building a fire against the freshly exposed stone until it cracked and turned brittle, and then pouring water over it, a laborious process that made the stone easier to break but produced clouds of suffocating smoke. An alternative method was 'plug and feather', which involved driving in a wedge and pouring on water so that the wedge swelled and split the rock. Whichever way it was done, it took many hours' work, all of it made possible by mutton fat, string and flint.

There were apocryphal tales in which miners trapped by a rockfall were rescued days later, having survived by eating their candles, supplemented by chewed-off bits of leather from a belt or the strap of a bag. But you would have to be desperate. Doug told me about one man, caught underground when the roof collapsed, who survived for two weeks on water dripping through the rocks piled precariously above his head. When he was rescued, starving and emaciated, he was asked why he hadn't eaten his tallow candle. He turned his head weakly and mumbled, 'Have you buggers ever tried one?'

Lead mines are much frequented by ghosts. At Goodluck mine, the Ghost of Gamaliel Hall wanders the tunnels, dressed in his woollen shroud. He died in the mine in the 1790s, either by hanging himself, or by falling down a mineshaft – accounts vary, but many sightings have been reported, most recently by a volunteer working on a bat survey. Miners frequently encountered the ghosts of their forebears, and 'T'Owd Man' was a spirit figure that must be respected and propitiated. Certain mines had a reputation for hauntings, and it could be hard to persuade men to work them. There were frequent reports of unearthly sounds heard deep underground – barking, wailing, whispering. In one cavern an explorer claimed to have seen a vast shape moving in the gloom, and claw prints a foot wide in the mud.

Every dark place has its ghost or its supernatural beast. But the thing that really makes me shudder is standing here at the bottom of a shaft looking up, and imagining what it was like to climb it at the end of a long day. The shaft is two feet in diameter, and there are stemples or bars of wood fixed at intervals, which would have been slippery with wet clay. The miner, bearing his bag of tools, climbed the stemples like the rungs of a ladder. It might be three hundred feet to the surface, and he climbed by candlelight, unless tiredness, mis-judgement or a sudden draught blew out the green candle and left him halfway up in the utter dark.

A ghost vocabulary haunts the mines too. 'Barmote', 'stow-blade', 'sole-tree', 'meer', 'freeing dish', 'jigging', 'fleak'. A 'gang' or vein of ore could be faked by 'glancing' and 'smith-oming' to trick a prospective buyer. A faulted vein was said to be 'squinted', or have 'slickensides' that made it awkward to follow. Every industry has its own jargon, but this particular vocabulary grew and flourished where vision was limited and survival depended on quick communication of complex technical information. It became part of the glue that bonded miners to one another, providing a way of describing their specific working conditions and methods, signifying belonging and cultivating a sense of community. It was a language spoken in the darkness of tunnel and level and sough, and now it exists only in books and museums.

Some things are only sayable in the dark. It might even play a role in the survival of a language. Some historians think that Welsh was kept alive in the coalmines at a time when it was being pushed to the brink of extinction. There were repeated attempts in the nineteenth century to suppress the use of Welsh and impose English as the common language; Welsh-speaking was castigated in schools, disallowed in official contexts and denigrated in public. But it lived on in private – in the home, in chapel and underground, where in the accommodating dark the miners were able to speak freely in their mother tongue.

Darkness is a refuge for the five distinct species of bat that swarm and roost in the Derbyshire lead mines, and for the rare tissue moth and fungus gnat, and the cave spider whose egg sacs I saw earlier, resembling bubbled plaster against the rock. It can be a refuge for us too, and we don't have to be underground to claim its sanctuary. In diverse times and places, it has provided opportunities for religious and political dissidents to share prohibited ideas and activities, and its hospitality also extends to personal encounters. We can speak more freely in the dark, and even use secret words that don't come easily in the light. In a bedroom at night, or round a bonfire subsiding into embers, the sheltering darkness allows a space for intimate exchange, confession and truth-telling, self-revelation and emotional vulnerability that would be impossible without its protection. I have been grateful for this so often in my own life, especially at times of stress and worry, when the way ahead was all slickensides, and no amount of smithoming could make it look viable.

Riverbank

In the fading light I study my postcard reproduction of the painting and try to match it up with the view in front of me. Tissington Spires – those dramatic limestone formations that resemble the towers of a fairytale castle – are unmistakable. But the trees are very different now, and in places distracting because of their apocalyptic appearance: at first glance it looks as if there has been a forest fire, but then I realise it's ash dieback that has left great swathes of the hillside covered with these blackened and broken skeletons.

I'm standing on the bank of the River Dove half an hour or so after sunset, trying to identify the very spot where Joseph Wright of Derby might have been one night in the summer of 1784. I imagine him standing and considering, assessing the possibilities, peering at his pocket-watch in the half-light and tipping his face to the ribbon of sky unfurled between the tops of the trees. Will the cloud break? Then unfolding his chair and taking porte-crayon and sketchbook from his satchel. It's a decade after he married the lead miner's daughter, heedless of family disapproval, and he's at the height of his powers. He is here to make the preliminary drawings for *Dovedale by Moonlight*, a painting I loved when I was a child. It shows a full moon hanging, watchful but dispassionate, over the gorge, at the point where the Spires thrust upwards from the steep hillside on the left. Clouds roll across the canvas, grey above but pale and gleaming where the moon touches them. In the foreground, branches extend their fingertips and the moon is delicately framed by foliage. Beneath, a bend of the river is brushed with light. There's a sense of stillness, as if everything lies under a charm; you can see it in the creaturely detail of trees and rocks, the silvery touches and the greenish tone of the whole composition.

The River Dove ran through my childhood, and I suppose I'm here on a kind of pilgrimage. I grew up close to Derby, where Joseph Wright was a local hero; we went on a school trip to the art gallery, and I was captivated by his pictures. I had no education in art and no sophistication in my way of looking, I just liked the way he captured light and dark on the canvas. This particular painting – the one in the postcard, its colours gradually draining in the dusk – felt personal in a way the others didn't, because Dovedale was such a familiar place, one my family returned to often on summer Saturdays and bank holidays, a place whose details I knew so well they felt like part of me. I knew there was a companion piece called *Dovedale by Sunlight*, but I wasn't interested in that, because it sounded exactly like a photograph pasted into our family album. In this one, though, the moonlight was transformative. I had never seen Dovedale looking like this.

Our tastes change as we grow. But even today when I look at the painting those greens and silvers and those unearthly rocks still lay a charm on the place, and ever since I first saw it I have wanted to come here after dark. I have revisited in daylight a handful of times in my adult life, always conscious that this is for me not an ordinary place like other places but a site of memory and legend, mythologised privately in my own head and reconstructed again and again in conversations with members of my family. One of the ways I can tell it has that special status of legend is that it's always sunlit, in memory and in photographs. The sunlight acts as a kind of preservative, keeping it ageless, fixing it in the realm of childhood.

On those more recent visits I experienced the place in a variety of other conditions. I saw it shut in by low cloud, and after torrential rain when the river roared and foamed. I saw it once in heavy snow, when icicles hung at the mouth of Raynard's Cave. Like everywhere else, I realised, it had its own modes and moods. As I knew from the painting, it must also have a night life, in which it was not *my* place at all, not an iconic landscape or a tourist attraction, not a set of snapshots in an album, but its own living, dreaming and mutable self. Now, looking back, I suspect I was thinking also about my parents: since their deaths I had become more aware of their lives – real and vivid and multidimensional, too big to be confined within the frame of my own life – and I wished I had understood this better while I had the chance.

I would go there at night, I decided, and encounter this dreaming Dovedale. No one goes to a place like this at night: it's a beauty spot, so what's the point of being there when you can't see anything? The question itself made it irresistible.

Joseph Wright made a great specialism of moonlight. In a letter to a friend he asks him to sell some pictures for him, 'the one moonlight, the other the effect of fire'. He names a buyer who might be interested, a Mr Cutler of Suffolk Lane: 'If he approves of the moon light he will take both ... but if he rejects it, pray give it house room till you hear further from me'. History does not relate whether Mr Cutler purchased on this occasion, but moonlight sold well, and when Wright needed money he would paint another.

The scene he evoked must also have stirred the hearts

and the pens of the Romantic poets, who were famously alert to the sense of awe such sights could provoke. The narrow gorge, striking rock formations and rushing river, all lying spellbound under a dramatic night sky, would have been irresistible. I'm reminded of a sonnet by Anna Seward we once had to copy from the blackboard for handwriting practice. I suppose it was chosen because of her local significance – she was born in Derbyshire, and lived in Lichfield most of her life – but it posed quite a challenge for the teacher as well as for us, being awkwardly strewn with dashes and apostrophes. The meaning was obscure to me, but I liked the opening line, 'Yon late but gleaming Moon, in hoary light', and especially the image of the moon resting 'on the cloud's dark fleece'. In my ten-year-old imagination I automatically set it here in Dovedale, because of Joseph Wright's painting.

I wonder whether he ever read this poem, and what he thought of such literary interpretations of his favourite subject. He was fascinated by the contrast between darkness and light of all kinds, and developed his own distinctive version of the style known as tenebrism. When in 1766 he exhibited his piece *A Philosopher Giving that Lecture on an Orrery, in which a Lamp is put in the Place of the Sun*, it was a sensation. The dramatic light effects emphasised the novelty of his subject: an orrery, or clockwork model of the solar system, being demonstrated in front of a small eager audience grouped closely around it. This state-of-the-art mechanical instrument brings home the fact that the sun is not a god or a magical entity but a source of light and heat, like a lamp. It shows how the planets – including our own – move around the sun, turning first towards the light and then away from it. A heliocentric view of the solar system had been established for over a hundred years, but seeing it operated in front of their eyes is a revelation for those gathered around the table. The tones and textures of skin are astonishing: the man on the left has a five o' clock shadow, the faces of the children are rosy smooth and the young woman is striped with shade and deep thought, as if realising the implications for faith, for meaning, for her sense of self. What, then, is really at the centre of things?

One of the first appearances of moonlight is in *A Philosopher by Lamplight* in 1769. The moon is just peeping from behind cloud in the top right-hand corner of the frame, indistinct and much less dramatic than the flame from the lamp. It seems to contribute nothing to the illumination of the man's face, although the river behind him is streaked with its light. It's given a more influential role in *The Alchymist*. From its vantage point 'on the cloud's dark fleece' it gazes down like an eye on the strange scene unfolding before us: the alchymist on his knees before his discovery in an attitude of worship, his eyes raised to the heavens. It was painted at a time when emerging science presented a challenge to established religion, and here the two are brought together in a relationship that is unstable but powerful. It's as if in these pictures we witness the Enlightenment taking place before our eyes: the faces of an audience emerging from shadow at the edges of a room, those of the boy apprentices radiant with new knowledge.

Anna Seward and Joseph Wright moved in the same circles, and were fellow seekers after knowledge. They were both involved in the Lunar Society, the gathering of scientists and intellectuals that was at the heart of the Midlands Enlightenment. Of course, women were not admitted to membership of the society – enlightenment had its limits, after all – but Anna was within its orbit, and the two of them were acquainted. Her poems, like his landscapes, are full of weather and drama, and often evoke night-time scenes, complete with moonlight. She must have noticed this shared preoccupation, because in 1788 she wrote to him to suggest a subject for him to paint, based on one of her own poems, 'Herva, at the Tomb of Argantyr'. His reply is polite and self-effacing; he concurs that the subject is an excellent one, 'but how to paint a flaming sword baffles my Art'. Instead of leaving it there, however, he begins to sketch out the possible scene: 'Would it strengthen or weaken the character, to lay it near the Sea upon a rising ground, and thro' an opening among the Trees low in the picture to see the Moon just rising above a troubled Sea?' He can never resist the attraction of moonlight. In one startling late piece, painted shortly after the death of his wife, a burning cottage blazes with emergency under the cool, dispassionate gaze of a full moon.

The car park closed at dusk, so we parked half a mile away at the pub and walked over the fields in the fading light. Then along the River Dove to the place where the valley broadens out and there, under the steep flank of Thorpe Cloud, was the familiar view of the famous stepping stones. Even in these conditions it seemed saturated with memories of childhood: long summer days with picnics and grandparents, walks where I trailed at the back complaining about nettles or gnat bites, the occasion when our gran set off bravely to cross the stepping stones, paused halfway to be photographed, then lost her footing and fell in. (The river was shallow and she was in no real danger, so she sat calmly in the water and waited to be rescued, holding her handbag above her head to keep it dry.)

The stepping stones are not the same as they were – the uneven limestone boulders were long ago capped to make them flatter and more regular. We negotiated them without incident, noting that we would have to cross back in the dark later, and continued northwards along the river. I felt a little electric charge of apprehension as the way narrowed and darkened ahead. In the deepening dusk it was beautiful but unnerving, that transition from one state to another. I could no longer rely on familiarity, and my tread was less confident. There were things to trip over and edges to fall down. Rocks shifted their expressions, water became moody and ambiguous. Ordinary movements were enigmatic, and sounds gradually divorced themselves from their source: water over rocks, bird in the undergrowth. I thought I heard those skeletal ash trees taking furtive steps towards us. Light zipped down my peripheral vision, and I blinked it away. I moved my head and there was another: a vertical shot like something with life and purpose of its own. I'm used to these flashes, especially at dusk, like a camera flash triggered automatically by the change in light conditions, tripping again and again when it isn't needed. It's vitreous detachment, the optician tells me, a normal and generally harmless condition that affects most of us over the age of fifty. The flashes will calm down, she says, as the eye settles and the brain adjusts to the new state of reality. But this evening there were dozens of these thin silvery spooks, each a little visible spike of adrenalin.

As dusk slid into night, all these apprehensions left me. Now in the mossy dark my surroundings feel familiar and friendly. I am a creature of the dark, I can be at home in it. The river gets dark long before the sky does, but is feathered white where it runs over rocks or fallen branches. The limestone steps give off a faint pale radiance and are easy to navigate. The crags lose their complexity and become two-dimensional, like stage flats rising up against the navy-blue sky. The narrowness of the gorge is exaggerated, and everything feels closer and warmer.

Tomorrow, when I visit the art gallery in Derby as I did on that school trip fifty years ago, I will read that Joseph Wright was not interested in painting generalised scenery, and disregarded the usual conventions around the picturesque, preferring just to paint what he saw. The 'seeing' was not entirely straightforward, however. I will learn that when asked about the process of making the picture, Wright admitted he did not in fact stand or sit on the bank of the Dove at night to sketch it under the moon. He considered such an excursion not only dangerous, but unnecessary. After all, he knew Dovedale, and he knew moonlight; all he needed to do, in the comfort of his own studio, was to put the two together.

So in a moment of disenchantment I will learn that I cannot possibly have been standing in the very spot where Joseph Wright stood, because that spot is a mirage. The best I will be able to say is that I stood in the place where I imagined he imagined he'd stood.

This kind of disclosure is an inevitable part of pilgrimage, in my experience. You set out hoping to find the exact place where something was made or thought or done, as if that bit of ground holds some sympathetic magic that can open a door between then and now and let you pass through. But you discover that the exact place has gone, or been built over, or it turns out that your hero was never actually there. The whole endeavour of pilgrimage relies on accepting that disappointment is inevitable.

Fortunately, there's another reason why I have come to Dovedale tonight: I'm hoping to see the ghost lights.

They were here thirty years ago, according to a young man called Oliver Rowlands who, like me, decided to visit Dovedale after dark. They may equally well have been here on that summer night in 1784, though Joseph Wright was not here to see them. Perhaps the majority of apparitions go unwitnessed, wasting their unearthly footsteps and flickerings on the empty places where no one happens to be around?

I stumbled across Oliver's story while checking a few practicalities in readiness for this nocturnal trip. I tapped 'Dovedale after dark' into the search bar, and there it was, on the website of folklorist Dr David Clarke: an account, over thirty years old, of mysterious lights dancing above the river. It didn't surprise me that they should turn up here, in a place already haunted by other kinds of ghost. I felt the pilgrimage expand to make room.

I got in touch with David, and he sent me a copy of the letter Oliver wrote to him in 1994, describing what he had seen. David, then a trainee journalist, had recently published a piece on the Longdendale Lights, an unexplained phenomenon in another part of the Peaks, and when Oliver read the piece he decided to tell his own story.

In the letter he says he drove here with a friend one spring night, and the two of them went for a walk by the river, intending to cross the stepping stones as we did earlier. They saw no one else on their walk. As they approached the stones, they saw two spherical white lights up ahead. The lights were intensely bright, and were moving silently in the air above the river in a kind of dance, tracking one another, moving in symmetry. 'They were moving up and down in crazy patterns far too quickly for them to be the lights of a motorbike,' Oliver explained. 'Anyway, there would be no chance of a vehicle of any type moving up and down the cliff so quickly or with such turns of speed. The experience, which lasted about three minutes, left us speechless. Then we started talking and questioned what it could be. There was something quite eerie about it. We eventually decided to turn towards the car again, and we were too frightened to look back, so we just kept walking and eventually broke into a run.'

A few months later he returned after dark with another friend, hoping but not really expecting that the same thing might happen again. This time they walked further along the river, well beyond the stepping stones and up the hillside, where they sat and looked down on the valley below. After a while a single white light, larger than the one he had seen the first time, drifted out across the river, rising and wobbling, before disappearing into the trees. 'To be honest,' he said, 'I thought that to see such an occurrence once would seem a miracle, but twice?'

Stories of strange dancing lights are of course common in folklore, and there's a lexicon of names for them: will-o'-the-wisp, Jack-o'-lantern, corpse-candle and so on. Isaac Newton – a near contemporary of Joseph Wright, and a fellow traveller of the Enlightenment – proposed that these mysterious sightings, which he called ignis fatuus or 'foolish fire', had a perfectly rational explanation: they were produced by marsh gas. 'Don't forget it was the Age of Reason,' said David when we talked about all this. 'Science was supposed to sweep away the old superstitions.' But the stories kept coming, and the sightings were often in areas far from any marsh, like here in the limestone valleys of the White Peak. The landscape refused to be disenchanted. Then in the 1980s a researcher at Leicester University carried out his own experiments and concluded that marsh gas was incapable of generating any such effect, even in the unlikely presence of a spark.

As a folklorist, David said, he doesn't try to solve the mystery of unexplained phenomena like ghost lights. The questions he investigates are about the stories themselves:

why people tell them, how they persist or change over time, how our interpretations of unusual experiences vary from one place to another and in shifting cultural contexts. 'It's often assumed that we're too clever these days to see a will-o'-the-wisp,' he said wryly. 'Of course people do still see them, but now they call them UFOs.' Lots of things shine at night, he added. No doubt I'd heard of the luminous owls. *Luminous owls?* Yes, he said, there had been sightings of them in various places for at least a hundred and fifty years. No one really knew what to make of it, though there was a theory that their feathers shone from contact with 'foxfire' or bioluminescent fungus growing in the holes in tree trunks where they roost.

I'm secretly hoping that by following in Oliver Rowlands's footsteps tonight I might see what he saw thirty years ago. In his letter he admits he found the experience frightening, but the intoxications of the dark would be nothing without a measure of fear, I remind myself, as I stand on the riverbank, looking back the way we've come, staring hard into nothing. And I've always known there's a special kind of glamour on this place. I'm just thinking about the word 'glamour', and how far it's travelled from its original meaning, when suddenly a bright white light emerges on the bend of the path and begins to oscillate towards me.

I'm simultaneously thrilled and appalled – for a moment it's as though I called the light and it came. I'm not sure I like having this much power. But as it approaches, I begin to have doubts. It's not quite as I'd imagined it from Oliver's account: not exactly spherical, more two-dimensional than that, and I hadn't expected it to bob along so close to the ground. Come to think of it, there's something about the regularity of the movement that looks familiar . . . oh, of course, it's a torch! A bubble of adrenaline pops and is replaced with a more mundane kind of unease. I can face any number of dancing ghost lights, but an unknown human in the dark is another matter. Then all the old anxieties come slithering in, a muddle of dangers real and outlandish. I actually consider climbing into a cave or crouching in the scrub to avoid the encounter.

Yet there's room among the fear for a thread of irritation. I had wanted to avoid all artificial lights on this journey, and had foresworn torch, mobile phone and camera flash. That way, I insisted, we could cultivate our night vision, which takes at least half an hour to adapt as the pigment rhodopsin builds up in the eye, and is easily destroyed by exposure to bright light, even for an instant. Now my rhodopsin level will collapse and my eyes will have to start the process all over again. By the time the torch is close enough for me to make out that the figure carrying it is a young man, I'm more annoyed than scared. He's sweet and disarming, though, asking whether we know a good place to pitch a tent. We chat for a moment before he turns back the way he came, and I blink helplessly in the sudden dark as if he's thrown dust in my eyes.

We start to think about heading back – we have quite a walk, and the stepping stones to cross, and it would be nice to get to the pub before last orders. I haven't seen what I was looking for, and I leave unsatisfied, wondering whether if we'd stayed just a bit longer the ghost lights might have put in an appearance, or the moon might have made it over the top of the hill. But unlike the kind of light we carry in our pockets, these are quixotic varieties that can't be switched on or conjured up at will, and that alone makes them worth searching for. Anyway, I did see Dovedale at night, and perhaps now it will slip the frame of childhood legend and become an actual place, just as my parents' lives, now they are dead, have slipped free of the scrap of ribbon, the trunk in the loft, the rooms full of family things that seemed to contain and define them.

As we climb the field to where the car is parked, there it is: a full moon, smudged by cloud, heavy and timeless. It's as though it's been waiting for us here. An owl hoots, and I feel the first stirring of a new obsession. I must come back another night and look for a luminous owl, perhaps focusing my search on some of those ancient oaks I saw this evening, where I might at least find traces of foxfire around a hole in a trunk . . .

When the pilgrim is thwarted, the pilgrimage takes a new direction. Towards the end of his life, in mourning and in poor health, Joseph Wright pushed at the very limits of tenebrism with a painting called *Romeo and Juliet: the Tomb Scene*. When the painting was exhibited it was badly received. It depicts a night when there is no moonlight to sweeten the scene as the lovers are drawn inexorably to their deaths. Where was the delightful radiance everyone had come to expect from him? What, no redemptive sliver of moon or comforting circle of candlelight? Afterwards he reflected that it was 'certainly painted too dark, sad emblems of my then gloomy mind'. His public agreed he had gone too far, but I think it's mesmerising. At the centre is the figure of Juliet kneeling by Romeo's body with her arm upraised, her spread fingers already spectral. Flowing over, behind and around her, lapping and overlapping shades of dark seem almost to lean towards abstraction, as if with a few more layers of paint this could become Kazimir Malevich's *Black Square*. But it is still very much a narrative work. Materialising across the huge slab of darkness is a still darker shape, the shadow of an approaching soldier bringing the decisive moment: *Yea, noise? Then I'll be brief.*

Frame

I am rehabilitating myself after meningitis by sorting through my father's old slides. This is a mammoth task, which has been waiting for just such an opportunity; they have been stored out of sight, out of mind at the back of a wardrobe since he died three years ago.

The slides were beautifully kept, in special boxes made of smoked grey plastic, with folded cards of annotations in his small tidy script. I am in my dressing gown, with the first of the boxes on the kitchen table. I prise open the lid, and it utters a reluctant creak.

My dad was a librarian, and his life at home as well as at work was structured on habits of organising, listing and cataloguing. He always kept a miniature notebook in his jacket pocket, the kind that came with a tiny pencil fitted into a sleeve in the spine, and whenever he noticed a job that needed doing around the house – a dripping tap, a stretch of guttering blocked by leaves – he would slide the pencil from its sleeve and make a note. The diary was for organising the future, and these folded cards with their neat captions were for organising the past.

To start with I tried to manage by holding each slide to the light and squinting, but they were so vague that way I wasn't always sure I had them the right way up. My eyes ached. I was at a loss to explain why I had chosen this particular project, when I was far from well and my visual processing was still so poor. Perhaps it was a test.

I ordered a handheld viewer on eBay, and now I have it the experience is transformed. Each click of the lever displays a scene from my childhood so brilliantly real I can only bear a few at a sitting. The houses we lived in, and the gardens where we played all day in summer, are shockingly small, and my parents unfeasibly young, but the clothes we are wearing, and the toys we are playing with, are as tangible as if I saw and handled them this morning. The tin trike made to look like a motorbike, that used to scratch your thigh every time you climbed on or off. The rusty swing with its black rubber seat. My brother's cowboy outfit, and my crow costume from the infant school play, which I had to throw off at a critical moment to reveal white wings and a sticky-out dress underneath: *Although I look just like a crow, really I am Fairy Snow!*

Then there are the others, the ones full of our absence. Views of distant hills, waterfalls, farmyards, boats and bridges. One a simple black square with nothing in it at all; he has marked the frame with an X, but I can't fathom why he kept it. Many with our mother, standing there for scale or human interest, and because he loved her and never tired of looking at her. Many more with no one in them at all, or just some passerby who glances into the lens, eats an apple or searches a rucksack. Hundreds, perhaps thousands like this. I will have to discard them, since none of us recognise the scenes they depict, and we need the space. But it's hard, when they speak so vividly of the years gone by, their colour and texture and interplay of light and dark.

In any case, there's no way I can bring myself to throw away these annotations. There's such eloquence in their simplicity and concision. Before I drop each slide into the box for disposal, I check it against the card and read the numbered caption aloud. Together they ache with meaningful insignificance, with the lived ordinariness of days in my parents' lives.

Zigzag path
Birds of prey
Our Christmas cactus
Very windy
John Cook's delphiniums
Another butterfly
Birthday dinner
Looks better in sunshine
At top of same road
Very cold walk
Evening view from our room
Filled-in quarry
Tame chaffinches
All this film affected by light

you were pale, too, almost luminously so in the moonlight

You were staying in a cabin in the woods in Michigan. You have a photograph taken on the grass in front of the cabin: a couple in their forties, with kind, open faces, and three children of indeterminate age, caught and blurred in the middle of some game. The mother had driven you, in their beaten-up old campervan, to a reservation where she had friends, and taught you a song on the way:

> *My paddle's keen and bright,*
> *Flashing with silver,*
> *Swift as the wild goose flight,*
> *Dip, dip, and swing.*

There was a canoe at the cabin, and someone suggested the three of you should take a dawn trip on the river the next day. It was still dark when you dragged the canoe over the shore and into the water. There had been a storm overnight, and the river was high and white in places. You didn't mention you had never been in a canoe. After an hour or so, passing through thick woodland on both sides, someone shouted 'Bear!', and you saw at the edge of your vision a black shape disappearing between trees.

On this same trip, you were with a crowd of teenagers at a place called Diamond Lake. It was night, and there were fire-flies over the water; you wondered whether they were the diamonds that gave the lake its name. Some of the other kids took off their clothes and ran into the water. You hung back as long as you could, unfamiliar with swimming in the dark and embarrassed by your own nakedness, but you couldn't bear to be the last to go in. Someone said the phrase 'skinny-dipping', and you knew as you peeled off your T-shirt and jeans that you were skinny. You were pale, too, almost luminously so in the moonlight, so that you wished it would go behind a cloud.

You struck out nervously towards the others, imagining huge fish and treacherous rafts of weed. Then suddenly everything changed. Now you were not you – awkward, unworldly, angular – but a body that moved in an element that you knew was water but that felt more like warm air, soft and spacious. You couldn't see it, except where the moon touched it with silver.

Dark lantern

Despite the same, scuffed moon
bolted at each corner to a listing pole

every intersection
offers a fresh start.

— Stuart Dybek, 'Night Walk'

Street

Certain things heard in childhood stay with us all our lives. They stick in the mind for reasons we can't always understand, changing shape as we grow and learn, so that we are able, many years later, to look back at them from a distance and understand them differently.

I was walking home at night with my mother, but I don't remember where we had been, or where we were when we stopped and looked at the sky together. We were talking about the constellations, and she was pointing out the Plough, and then the Seven Sisters. She told me there were in fact more than seven, and I said I could only count five, and she sighed and said 'Stars aren't as bright as they used to be.'

I was troubled by this remark, and the sigh that accompanied it. She was sad, which I always found frightening. It seemed such a mysterious thing to say, and I was still puzzling over it in bed that night. Could it really be true that the stars were fading, wearing out like a torch beam when the batteries ran low?

One night in London a while ago I was sitting at my desk with an essay on the screen in front of me. I wasn't actually reading it; it was late and my mind was drowsy and blank. After a drifting minute or two, the monitor went into sleep mode, the electronic glare shut off abruptly and the room, lit only by a table lamp in the corner, felt suddenly softer and warmer. My attention flicked to the window, and the night beyond the glass.

Next door they had installed new lights, and a timer that switched them on at dusk. One, affixed to the wall of their house, illuminated the concrete paving, the bare winter branches of plants in pots and tubs, a dustbin, two folding wooden chairs. They were so bright I could see beads of rainwater on the arms of the chairs.

Light slid over the fence and made a stripe, sharply delineated as a path, along one edge of the actual path, but the rest of our garden lay in shadow. As I watched, a vixen jumped down from the fence, looked around a moment, then crossed from left to right, from the light into the dark. I couldn't see her any more, but I heard a thump as she leapt and landed on the roof of the shed.

I wondered about the mosaic of spaces she was travelling through that night, some lit and others not, and how it felt to her to move between them. Foxes are nocturnal by nature, but she didn't seem bothered by the light that fell on her as she moved from garden to garden, looking for scraps. Urban foxes have adapted, and you often see them out during the day. They are not shy like their country cousins; they notice your approach but continue their task of scavenging from a torn rubbish bag, breaking off reluctantly if you get too close, and looking you evenly in the eye before loping away in no particular hurry to slink under a hedge or between buildings. Who knows how their days and nights are patterned, these feral creatures born in the city like so many generations before them.

We humans have also got used to changes over recent decades: bright light, cheap and plentiful, tempering the night indoors and out, in rural as well as urban areas. It pours from streetlamps: intense, bone-white, bleaching the pavements, walls and hedges, and giving every upright thing a shadow. These powerful lamps, often poorly positioned and inadequately shielded, don't necessarily make our streets safer; they tend to create excessive contrast, making the dark outside their ambit even darker. For the driver, hazard perception becomes more difficult; it can be like coming out of the cinema on a bright afternoon, dazzled in the sudden daylight.

Walking past a 1960s estate a few nights later, I noticed that some of the roads between its residential blocks and gardens had hung on to their amber lighting, and it felt for a moment like time travel. We had them on the estate where I grew up, and we used to say they ran on orange squash. Colours all looked very similar under their light: it collapsed the whole spectrum to shades of sepia. It stained the leaves of any tree that grew nearby, and playing out on a winter afternoon we would jump up and try to tear one off, but whenever we managed to get one it was ordinary green. We knew it would be, really, but there was always the chance of magic.

They seemed to have some darkness mixed into their light, those sodium lamps. I once had a nightmare in which they were mouths that sucked in shadows and spat them out. The shadows ran and pooled on the cinder path, alongside the wall where I'd once seen a man stand swaying and grunting, and made their own dark rivulets that dried to a stain by morning. On the way to school, of course, it seemed ridiculous. Who at that hour of the scrubbed face could really believe in those hungry mouths, jammed open and feeding on shadows?

Our cities and towns look very different, now the smudgy romance of the amber lamps is gone. Their colour rendering was poor, but it was a warm light and it didn't dazzle. 'Its

poor quality was its beautiful quality,' said Kerem Asforoglu when we met for coffee near his office in north London.
Kerem, a lighting designer and dark sky activist, is at the
forefront of a movement to make our lighting practices more
aesthetically pleasing and more sustainable. The LED was
adopted for its greater efficiency, he told me: it delivers many
more lumens per watt, production costs are much lower and
units last longer. Warm tones are available, but it's the hyper-
efficient blue white that predominates because it's brighter
and cheaper. It's made light pollution much worse, he said,
and anyway humans generally don't need to see colours
at night; we have evolved with scotopic vision, where the
eyes adapt naturally to low light, and this has served us well
until now.

Old and new jostle against each other: a few yards from
the amber streetlamps two new residential blocks were being
built, and on the hoarding was a single unshielded fitting so
aggressively bright it hurt my eyes. It made the construction
site look like a stage set for a melodrama, the clutter of signage
hinting at the narrative about to unfold: DEEP EXCAVATIONS,
NO IDLING, a grotesquely oversized hand with a blood-red
diagonal line through it. Light was pushed up the girders and
plastic sheeting, up the walls and windows of the neighbouring blocks, up the mast of a tower crane, all the way to the
shining cab with its levers and dashboard, and out along the
jib where a stack of concrete blocks rested improbably on a
little platform, like a bad plot device.

Down a narrow alley between buildings, behind chained
and locked gates, darkness was trying to take hold, but it was
policed by a security light that had it kettled in the far corner.
Light can be a form of surveillance, a means of patrolling nocturnal activities. The bohemian and hedonistic pleasures of

the night were once enjoyed under cover of darkness, and the introduction of street lighting in seventeenth-century cities was so strongly resented that people took to lantern smashing. But now our nightlife depends on light as much as dark. A mile or two from here, I thought, in the city centre, squares and streets will be garish and giddy, light running together like coloured paints under the tap. Streetlamps, traffic lights, the lit windows of buses – these traditional sources of light will barely be noticeable in the flood that gushes from theatres and bars and restaurants over the wet pavements. Mannequins in the brilliant windows of clothes shops, pale and crazed with insomnia. Digital billboards flickering and flaunting, and quaint flashing neon promising PEEP SHOW, TABLE DANCING, GIRLS GIRLS!

The city has more than one kind of nightlife. It's a place of hectic invention and reinvention, where the shape of the skyline is never settled. Streets are drilled for the installation of broadband or for the patching of Victorian sewers, and diverted traffic surges along the surrounding roads. Bars and fast-food joints fall vacant then reopen, removal boxes are carried in and out of flats, people with nowhere to go seek shelter in doorways and railway arches. Bins are emptied, street sweepers vacuum the litter and dirt from the gutters. All this activity spills over from day into night, blurring the edges between them. It's serviced by night buses and all-night cafes and volunteers on soup runs, all superintended by stern white light, and invigilated by cameras bolted to the lamp-posts.

The streetlamp outside my hotel room – so bright it seems to scorch its way through the curtains – is hung with a huge cobweb that fans out under the white light. I have no idea how long it's been there, but I wish I had been able to witness the process of construction. It looks like a pane of glass in a shop window which has been bashed with an implement, but not quite hard enough to break it right through, leaving it crazed with silvery fault-lines, unexpectedly beautiful.

Is this a hospitable dwelling place for a spider, I wonder? I think I've read that most spiders don't have great eyesight, so perhaps the coming and going of the strong light doesn't bother her. Of course, she has built her trap here because this is where the flies and moths gather, compelled by the light. They can't resist. They circle and circle until they touch the web and are caught fast. It's estimated that a third of them are killed, either as prey, in collision with the lamp, or simply from exhaustion. I remember a camping trip in my youth when I reached to switch off a battery-operated lantern and found my whole hand and wrist sleeved in insects. Imagine this on a city street with a streetlamp every few yards and no one coming to switch them off.

Why are insects attracted to light? People must have been asking this question ever since they began to sit around fires and carry blazing torches. It's such a well-known phenomenon that we've encoded it in metaphor, as a way of describing self-destructive behaviour, the tendency of an individual to be drawn irresistibly towards the very person or thing that could destroy them: *like a moth to a flame*.

More than half of all insects are nocturnal, and it's been known for a while that some of them use celestial objects in navigation. The dung beetle or scarab, for instance, consults the Milky Way to keep it on a straight course as it rolls its ball of animal dung across the desert sand to its hiding place. Perhaps this was the answer to the age-old question. Could it be that artificial light sources, such as streetlamps or city buildings, were mistaken for the moon, provoking a sharp change of course which took the insect in the wrong direction?

In 2023 a research project found that a very different mechanism was at work. The researchers used new technology – battery-operated high-speed cameras and motion-capture processes – to film moths and other insects in an attempt to understand this eternal mystery. The results of the study contradicted the longstanding 'lunar hypothesis'. The surprising conclusion was that insects are not attracted to streetlamps at all.

Insects don't 'feel' gravity in the way we do, because their bodies are airborne rather than being in contact with the ground. Instead, in order to be sure which way is up, they make use of a reflex called the dorsal light response. This keeps them orientated with their backs turned to the sky, which is generally the dominant source of light. The new footage shows that as a moth flies in range of a lamp, the reflex is triggered and it turns its back to the light. Instead of orientating itself against the flat of the sky, it now ends up circling with its back to the lamp, trapped in an orbit it never intended to enter.

Dr Sam Fabian, one of the team of researchers on the study, told me he had always been puzzled by the lunar hypothesis, which didn't seem to add up. He'd been troubled, too, by the implication that insects were a bit thick. He has been studying them for years and knows they are intelligent and complex creatures, with ways of perceiving the world that we can't even imagine. I told him about the boy on our estate who said moths thought the streetlamp was the moon, and he laughed. 'I always felt it was unfair of us to suggest they were stupid for having a light or lunar fixation,' he said. 'It seemed more likely we were missing something.'

Gazing at the cobweb as it trembles under the light, I think of something Paul Evans wrote in *How to See Nature*. Standing under a streetlamp, he observed that 'the shimmering pixels of insect wings can be seen as flashing dots of synapses and neurons'. Even with our big brains, we don't always get it right. We are all like moths to a flame at times. Not hurling ourselves into the fire to be consumed, no, not that after all. But desperately trying to stay the right way up in a world where things are not as they should be.

Recently a new word was coined to express the feelings of grief and lament over our disappearing night skies. 'Noctalgia' is generally translated as 'sky grief', though I prefer to call it 'night grief', a more capacious term that speaks of a general bereavement: the loss of night's essential quality of darkness, and the sense of impoverishment and sorrow we feel as a result.

But it's the disappearing stars that hurt most. Another recent coining, 'solastalgia', refers to a feeling of homesickness experienced not when you are away from home but when your home environment is changing in distressing ways. The home environment of our shared skies has changed so much already, and is threatened as never before by over-illumination, skyglow and satellites.

Distress builds and builds until language has to take notice. Words like noctalgia and solastalgia take root because we need them urgently to describe our responses to environmental crisis: grief for what has already been lost, and anxiety about future losses. At times like this, we can't afford the luxury of mute despair. We need new vocabulary in which to articulate the emotional impact, so that it becomes spoken, heard and shared. Mere nostalgia is not enough; we look for a rallying call. When we can talk, we can act.

Of all our environmental problems, light pollution should be the easiest to solve. All it requires is dimmer lights, and fewer of them, and for them to be switched off when they're not needed. Where this has been achieved, the effect is immediate. It saves money, and leaves no residual mess to clear up: as dark skies advocate Dani Robertson writes in *All Through the Night*: 'There's no rattling around the planet for thousands of years like plastics, no slicks to be mopped up like oil spills.' But as Dani knows, through her own pioneering work with Kerem and others, success depends on altering public perceptions. We have to re-educate ourselves and each other, become accustomed again to lower light, used in moderation. Trying to erase the dark, to blast it out of existence, is not only damaging but futile.

Villes éteintes, a series of images by digital photographer Thierry Cohen, offers a fantasy of how our city skies would look without any light pollution at all. The images are composites, each a photograph of a city which Cohen has brought together with one taken in a remote location like a desert or wilderness. He matched two locations on the same latitude, so that the configuration of the stars would be the same. The resulting pictures are both beautiful and unsettling. Here's the City of London skyline with its familiar skyscrapers – the Gherkin, the Cheese Grater, the Walkie-Talkie – and there are their shadows on the oil-black Thames. The shadows are cast by the night sky, which towers with light. The buildings themselves, the streets around them, the whole environment, are dark. No lit office windows, no streetlamps, no screens or neon or traffic lights.

These images are playful, a trompe l'oeil. But they have a starker aspect too. For the time we spend looking at them, the city itself – excessive, hubristic, bloated with light – is '*éteinte*', or extinguished, and in that moment we are confronted with the truth: it is *nothing* in comparison with that sky, with the vastness of space and the billions of stars around us.

The word '*éteinte*' can also mean 'extinct', and although in our arrogance we might tend to think of the night sky like an endangered species facing extinction, it will of course outlast all our cities.

We should speak not of darkness but of darknesses. Like water, it differs from one place to the next. All water is made of the same constituents, two parts hydrogen to one part oxygen, but each body of water – lake, river, ocean – has its own character, which itself changes from one shore to another. Urban darkness has its own partial, intermittent and shifting character, and its own distinctive repertoire of pleasures and disquiets. It doesn't find you out; you have to search for it. It doesn't occupy the sky and fill the streets as it does the lanes and fields around our cottage. It's more fragmentary; it manifests as shadows, pools, tunnels and hinterlands the artificial light can't quite touch.

That unshielded lamp on the building site hoarding spilt across the street and lent an unearthly glow to a single clump of ragwort growing by iron railings. Beyond the railings a park was locked and dreaming, breathing out a scent of lime trees in blossom. It was a fitful sleep; light washed in from the street on all sides, eating away at the edges of the dream. I stopped and peered in. There were a few people in there even though it was hours since closing time. There is a mildly transgressive thrill in being in a park at night, having the run of the place, following paths between bushes that seem wilder and more interesting in the dark, sitting smoking on the vacant benches, watching a fox cross the playground or the tennis courts. I could make out three figures moving soundlessly: one on a bike, crossing very slowly left to right, and two clasped together, either in an embrace or in a fight to the death – I couldn't be sure which.

Further along the road, a concert had just finished in the old church. The stained-glass windows would once have dazzled with the brilliance of their blues and reds and golds, but city lighting now dimmed their colours and subdued their religious messages. Light leaked from the open doorway, revealing cool sepulchral shadows between the yews and the broken tombs. I spent a lot of time there when graveyards were my subject, and I knew how darkness accrued in its interstitial spaces at night, where foxes dug their dens and moths sipped at the ivy flowers. But the churchyard shadows were thin in comparison with the franker dark of the garden cemetery along the street, where the dissenting dead were left to their inscrutable sleep. It was cool and inviting in there, out of the city glare. A heavy chain secured the gates, and I was surprised such a simple mechanism could hold so much dark. I could read the inscriptions on the nearest monuments, but the path soon slipped away into invisibility.

In an age of absolutism, we should value every bit of darkness we have, I thought We need it to protect uncertainty, and to remind us that there is more to the world than its lit surfaces beaming back our own reflection. I crossed the road and looked in at the shop windows. The estate agent had left the display of particulars uplit in tones of green and mauve, but other businesses were truly shut for the night. In the barber's shop I made out a long row of leatherette chairs, each slung with a nylon cape ready for the next day; it was impossible to tell how long the row was, and whether there was a mirror multiplying the chairs. Next door, in the impenetrable dark of the charity shop, the racks of clothes receded into the historic gloom. But it was at the shuttered Chinese restaurant that I encountered the deepest dark of all. The place had been closed for years, darkness accumulating and condensing behind the steel and glass. I peered through holes in the metal screen but could make out nothing. Were the chipboard tables with their red cloths still there? Did the lucky cat with its gold coin still stand on the counter and beckon?

Just as life turns up in the most inhospitable places – wildflowers colonising the cracks between paving stones and rooting in the dust around the windows of office blocks – so there are dark places in the sterile brightness of the city. Pockets of dark that are often overlooked, like small fish surviving unnoticed in a polluted stream. Park, graveyard, alley, railway arch; and the inner spaces where it retreats for the night behind shutters and toughened glass. Our own domestic rooms, their doors closed to keep out the brilliance of stairwell and corridor, thick curtains drawn against the gaze of the street.

As I walked, I looked up at the windows of the tower blocks, some lit and others already dark for the night. I was trying to get to the heart of something – a prickle of words, a thought, something about being alone in bed at night, and in the privacy of darkness and solitude feeling your life coursing through you, recognising the shape and texture of your own past and the things that have mattered to you. I was concentrating hard, trying to let it in, and the wine I drank earlier was opening up a space for it, and I felt a poem begin to unfurl itself in my head –

At night in the house
   a river runs through her

     carrying its burdens
  the golden barges   the dead griefs and the quick fishes

I imagined it like a long strip of paper unrolling from some kind of spool within me, where it had been tightly curled for a while. It doesn't happen often, a whole poem unrolling in one go. Usually it's bits I have to pull at, tear off and piece together.

  the scrolled barges
  with their forgotten cargoes
  of sugar   tobacco   raw silk

I walked faster, to get back to my room and write it down before that paper strip could be blown away down the street and over a wall and lost.

Fragments of music floated over the traffic noise, blotted out for a moment by a siren, then cut through by the bleeping of the pedestrian crossing, then floating free again. A man stood alone in the doorway of the chippie, playing a mouth organ as if blowing a small fire alight in his hands.

    and the illicit little night boats
      tied up swiftly
    while the moon was behind a cloud

A few more bars, and he pocketed the mouth organ and strolled inside for his tray of chips. Sparks of tune flickered into the air and were lost in the luminous fog of sky glow.

Cell

Brian Keenan, who was held hostage in Beirut for four and a half years, has spoken and written extensively about the experience. He spent much of the time blindfolded in a windowless cell, and there were power cuts which left him for many days with no source of light at all.

During one of these periods he had hallucinations which involved not only vision but hearing and all his other senses. 'I remember on one occasion waking up,' he recalled, 'and having to squeeze my face and my chest and thinking to myself "am I still alive?" There was nothing there to confirm to me that there was human existence outside me or even in me.'

Throughout history, enforced darkness has been used as a form of torture, from the oubliette, where the medieval prisoner was simply dropped into the dark and left to die, to the 'enhanced interrogation suite' used by the CIA at Guantánamo. Deprivation of light is often part of a more comprehensive package of sensory deprivation, which at low cost quickly provokes extreme hallucinations followed by psychotic breakdown.

An equally efficient alternative is deprivation of darkness. Here the detainee is kept in a completely white room, constantly lit in such a way as to eliminate all shadows. Like the cell where Winston is incarcerated in *Nineteen Eighty-Four*, it's a place of brainwashing, disorientation and madness.

In either case, the definitional dial is turned up far higher than normal. In the torture business, light is light and dark is dark: there is no room for any shadowy middle ground. It must require great commitment and dedication to stick with such a totalising mode of thought. But perhaps it's not as hard as I think, especially in this age of certainty with its border walls and thought crimes. Ambiguity would make the job impossible from the start; the heart must already have been hardened, empathy torn out by the roots and doubts torched like so many invasive weeds.

Museum

In 1948 an exhibition was held at the Science Museum to mark one hundred years of electric lighting. The exhibition began with a section titled 'Battle with the Dark' and ended with 'Triumph over Darkness'. There was even an exhibit called 'Permanent Daylight'. These titles, this language of conquest and subjugation, is hair-raising to the twenty-first century reader. In the archives I pored over the exhibition brochure, but found no recognition of the value or importance of darkness, or the faintest squeak of concern about its loss. It is not important, neither is it beautiful; it is simply a problem to be eradicated. In a tone of breezy postwar optimism the brochure tells of 'the surge of progress that now presages a future freed from so many of the gloomy restrictions that forced our forebears to leave so many labours undone'.

The mention of labours is certainly at the heart of the matter. The availability of artificial light was one of the great drivers of the second half of the industrial revolution, allowing work to continue around the clock and abolishing the need for the machines to go quiet overnight. And the exhibition brought the visitor right up to date with a series of photographs of bare-chested miners kneeling on the ground and cutting the coalface with picks. They were the guinea pigs testing an experimental installation of the latest fluorescent lighting which, the text enthused, was good for the health and practically indistinguishable from sunlight. This revolutionary new technology would offer mankind 'his first opportunity of bringing real light to the night sky, permanent daylight to factory, office and shop'. Now there would be no need for labour, production, bureaucracy or transaction to pause at night. There would be no need for night.

No one in 1948 could have foreseen the current pandemic of artificial light; the term 'light pollution' hadn't been coined yet, and there was little or no awareness of harm to the environment or to human health. But the language of battle and triumph echoes ominously down the years to the present, like a warning heard too late. The surge of progress has gone on and on.

I'm not particularly acquisitive, but occasionally there is an object I long to own, for reasons I can't quite fathom. These longings arise out of whatever writing I'm doing, and the object in question is imbued with special significance because of its association with that writing, or more accurately with the feeling of intense engagement – obsession, I might call it – that is such a key ingredient in the writing.

Sometimes the object is not longed for, but arrives by chance or will of its own. So it was with the headless plaster statuette of the Virgin Mary that stands on a shelf near my desk. We were walking through a street market in Toulouse and I spotted her in a box with a lot of broken stuff. I recognised her from a poem I'd written not long before, in which a decapitated plaster virgin speaks about her worldly longings: 'a little shrine by a spring, and red stain for my lips'. Only after we had gone round a corner and crossed the road did I realise that this was a significant meeting, and not one I should refuse. I went back and asked how much. The man behind the table looked at me as if I was mad. 'But she has no head,' he said, tapping his own head in sympathy, as if afraid I might not have noticed. Eventually we agreed on two euros and I brought her home in my suitcase, wrapped in my swimming towel.

Recently I have been coveting lamps and lanterns. Safety lamps used in mines, oil lamps in people's homes, magic lanterns, like the one an enthusiast brought to a Puffin Club picnic I attended in 1972 – I have no memory of the show at all, but I remember the excitement of sitting in a darkened room with lots of kids I didn't know, and the devoted way in which he showed us that Victorian machine afterwards, all brass and black wood, with a little sliding frame for the pictures.

But the kind I would most like to find at a street market or car boot sale is a 'dark lantern', a kind of early flashlight, a simple but ingenious piece of equipment which allowed the light to be shut off without extinguishing the flame. It had a kind of fluted cap on top which allowed air through so that the wick would continue to burn, and a shutter you could slide round inside the glass. If I spotted one of these in a box of broken stuff I would certainly turn back for it, haggle if necessary, insist even if the cap was missing, bring it home and stand it on the shelf next to the headless Virgin, in my small domestic museum of obsessions.

In fact the only one I have seen so far was in a display case in a museum, where the label told me that in the nineteenth century the dark lantern, sometimes known as the bull's-eye lantern, was a standard piece of equipment for watchmen, policemen and soldiers. This was mildly interesting, but got nowhere near the heart of its appeal, which as far as I'm concerned is less utilitarian and more aesthetic. I'm like the boy in

Robert Louis Stevenson's essay 'The Lantern Bearers', longing to get his hands on one of these thrilling objects for a few hours. Stevenson describes the bliss of having one hidden under his coat, ready to be brought out and operated when the moment was right. 'The essence of this bliss,' he writes, 'was to walk by yourself in the black night, the slide shut, the top-coat buttoned, not a ray escaping, whether to conduct your footsteps or to make your glory public,—a mere pillar of darkness in the dark; and all the while, deep down in the privacy of your fool's heart, to know you had a bull's-eye at your belt, and to exult and sing over the knowledge.' In the dark, he and his friends recognise one another less by sight than by the distinctive smell of hot metal, and gradually they congregate in some hidden spot, where each boy slides the shutter to open up his light, and together they 'delight themselves with inappropriate talk'. Though I've never yet held a dark lantern, let alone tucked one inside my coat, I feel I too can smell the blistering tin and imagine the joy of possessing such an object, even if only for an hour or two.

Stevenson was drawing on childhood memory when he wrote about the dark lantern, but the practical realities of lighting hadn't changed much; oil and wick were used throughout his lifetime as they had been for centuries. 'The Lantern Bearers' is not about nostalgia for a vanished technology, but an impassioned claim for the existence of the soul. Not that he uses that word, with its religious connotations – he rejected the Christian faith as a young man, and remained agnostic for most of his life – but he draws a parallel between the burning lantern hidden under the boy's coat and the 'golden chamber' at the centre of every human life, no matter how dull and stolid that life might appear to others. The opening and closing of the shutter across the flame is not a battle between light and dark; there is no question of triumph. Imagination, feeling, selfhood, creativity – each of us carries this fire within us, revealing and concealing it when we choose, balancing outwardness and interiority, negotiating between our public and private selves.

Museum artefact, golden chamber, object of desire . . . however I think of the dark lantern, its greatest charm is its name, which sounds like a lovely paradox, a wishful contradiction in terms. A lantern that sheds darkness! If only I had one to take with me into places where the light is fierce and relentless, and use to shed some cool, restorative shade on things. This would not be something to confine to a museum shelf. I would place it on the bedside table in this hotel room, where it would cast a sweet pool of dark to sleep in. I would point it upwards into the orange wash of a city sky and see the Milky Way, or even hang it from a streetlamp so that insects could right themselves by moonlight and starlight, just as they did for millions of years before we came along.

Stage

Poets spend a lot of time ruminating on the things that go badly at readings, and I'm no exception. It starts with the crap microphone that kept cutting in and out. The person on the front row who fell asleep. The lack of a table where I could put down my glass of water. The lack of a glass of water. After a while it moves into more uncomfortable territory: the stupid things I said, the times I tripped on a word or missed it out altogether. The bad choice of poem. *The bad poem.*

What a strange thing for a self-conscious person to do: draw something up from a deep place inside yourself, write it down, publish it and go round reading it out in public. So many opportunities to be humiliated! Perhaps it's a kind of regression therapy. I only have to imagine a stifled yawn, or the scornful curl of a lip, and there I am, fourteen years old again and wearing the wrong skirt.

One of the things that varies from venue to venue is the lighting. No one gives much thought to it in libraries and schools and universities, and readings in pubs are often so dim you can't see the page, but those in theatres have proper rigs and technicians, and you can find yourself spotlit and staring into the void. A cough, a shuffle, a phone vibrating in someone's pocket – these are the only clues that there is really anyone out there. Eye contact is impossible. The light bounces off your varifocals, and exposes you in ways you hadn't anticipated, its merciless glare showing up your pasty skin, your dry mouth, the nervous tremor in your hand. At least, I'm imagining it does, as I stand alone on the stage, blinking into the dark as if imploring it to take pity on me.

Once I settle in, though, say somewhere around the third poem, the dark starts to feel fine, and then better than fine. A kind of recklessness takes hold – I can say what I like into that ineffable space! I forget that I'm lit up like a historic monument. Now the tolerant darkness seems to cover me too, and I can almost kid myself I'm invisible. OK, you can see me, but you can't see me seeing you see me.

Train

It's already dark as we run through the suburbs and into open countryside. It feels as if we have floated free of the tracks and are drifting, loose, slippery and unattached. We could go anywhere, turn in any direction, loop, sway and tilt in this undefined space. The train seems to navigate for herself, forging her own path, brave and unpredictable. We nose on into the night, and I feel touched by the courage of the endeavour. If all goes well, we will come in time to a place where there are lights to guide us home.

Later we enter a tunnel, where on the other side of the window there is nothing but black. Then flash – a short bar of white light. Flash – another. They domesticate the darkness, raising in the imagination a piece of ordinary labour: a workman on a ladder, the squeal of a drill, the bulb screwed in and the plastic shield tapped into place. But there must have been a light to make it possible for him to install the light.

The mental contortions we make to distract ourselves from the terror of knowing that we are travelling under the sea! Above us is an immense volume of water – how would I begin to calculate it? I start to add up all the things the water might hold. Ferries full of passengers, cargo vessels stacked with containers full of consumer goods and construction materials. Kelp beds, vast schools of fish and pulsing swarms of jellies. Fibre optic cables, and many dead cables of earlier vintage. The eternal cross-channel swimmer in his coat of mutton fat. Even the seabed is above us, though I know it as the very bottom of things, the place where everything comes to rest. I have seen it on film, littered with skeletons and sunken wrecks.

And all of that abyssal dark lies under the living dark of night, and night under the unfathomable dark of space. Dark is stacked on dark, and our train is just the thinnest streak of light beneath.

Is it really possible to pass underneath all this? I lean close to the black square of glass and open my eyes wide and almost believe it – belief is like a taste on the tongue, faint and fleeting – but then the bar of light flashes across and I recognise it: just a corridor, like all the other corridors in the world. And in any case, our carriage is lit, we have magazines and croissants, there are power sockets under the seat in front. Just then, with a sound like air sucked violently from a tube, we are out – and we wake, and it was all a dream –

the moon looked back at you without flinching

You went to three doctors.

The first said, 'Sounds like boyfriend trouble to me.'

The second said, 'It's only a small war, it'll soon blow over.'

The third said, 'Take this four times a day and watch out for dizziness drowsiness confusion nausea headache mood-swing seizure blurred vision slurred speech insomnia nightmare paranoia psychosis.'

You hid them at the back of a drawer and waited till it was dark. You closed the curtains, then you tipped out the bottle and looked at them. Your mouth was dry. They were quite a handful.

You put them all back and shut the drawer. Then you pulled back the curtains and looked at the moon for a while.

The moon looked back at you without flinching. You could feel them working already.

North

Darkness will not erase you the way it erases day with night
because darkness is not the clock but merely the time
falling away from the clock's circular face.

— Alice B. Fogel, 'Forgiving the Darkness'

Island

As we fly along the west coast of Norway, with its fjords and many snowbound islands, we outrun the light. The sky behind us and on our far left is still translucent, the horizon lipstick pink, but ahead it lours grey and purple as if a storm is gathering. So, I think, darkness is a place – you can travel into it. The voices and devices of the aeroplane cabin recede; all that matters is what I can see at this small window. If we reach a threshold between day and night, I want to be conscious of it, to feel myself cross from one state to the other. I watch intently as light is pulled slowly under the horizon like a trailing garment, the last part of a gradual and dignified withdrawal.

But of course there is no threshold. Day and night, light and dark – these are not binaries, though our language keeps pushing us to think of them that way. Instead we plough into a huge bank of cloud, and the shapes of the hills and lakes beneath us lose their definition and melt away entirely. It's several minutes before we emerge into the clear again, and by then we are in a kind of twilight, a trembling hesitation between states.

My destination is polar night, or at least a version of it; this is not a single phenomenon, but is highly variable depending on where you are. At the north pole, there's only one sunrise a year, and over the archipelago of Svalbard there is continuous darkness between late October and the middle of February. I have been fascinated by Svalbard, and its northern coast in particular, ever since I read the work of Christiane Ritter, who spent a winter there in the 1930s. But Svalbard has one of the most fragile ecosystems in the world, and I couldn't justify flying in for a few days. In any case, I have people to meet in Tromsø, and the email address for someone who can arrange accommodation on a small island out in the north Atlantic. It's a thousand kilometres further from the north pole than Svalbard, in the 'civil twilight' zone, where although the sun does not rise above the horizon at this time of year it gets close enough to afford some muted illumination each day, in what's sometimes called the 'blue hour'. But the rest of the time it will be dark.

Friends were incredulous when they heard we were jetting off to northern Norway in January. 'You've got your compass the wrong way up!' said one, to much hilarity. A British January can seem gloomy enough, but even in the most overcast weather the sun rises each morning and does not set until late afternoon. Yes, we could have gone to the Canary Islands for the same money. But this is a chance to immerse myself in darkness, and find out how it affects all life in the Arctic, human and more than human.

Like so many of my favourite books, Christiane Ritter's *A Woman in the Polar Night* turned up in my life by accident. I was skimming a page of reviews – neglected non-fiction by women writers, or something like that. A new edition had just come out, and the reviewer described it as a classic. I wondered whether I ought to be embarrassed never to have heard of it.

It's a sensational story. At the suggestion of her husband,

Christiane travelled from her home in Austria to spend a winter with him on Spitsbergen, now Svalbard. For reasons that are never made clear in the book, he had for some time lived away from home in the Arctic, interspersing research of some kind with hunting and trapping. He had described the place in his letters home, and Christiane had responded by saying how exciting it sounded. Well, why didn't she come and join him and see it for herself?

It was a shocking and formative experience. Either her husband had grossly misrepresented the living conditions, or she had somehow omitted to ask him for details. On arrival she was aghast to find that her home for best part of a year would be a tiny ramshackle hut on a bare, rocky shoreline known as Grey Hook, several days' journey from the nearest neighbouring settlement. Mysteriously, he had also failed to mention that they would be sharing this hut with his friend Karl.

The marriage itself is enigmatic. Its currents and dynamics, distances and intimacies are no more than shadows flitting across the text. The reader is left to speculate, because this is absolutely not her subject. Instead she is intent on recording and describing the life around her – the seals and polar bears and seabirds she hunts, eats, observes and avoids – in that edge-of-the-world place. She writes, too, about her own routines, the continual business of getting food and fuel to stay alive. In doing so she evokes an experience of remoteness and solitude so extreme it feels like a work of fiction. The two men are often away hunting for weeks at a time, and during those periods she lives in complete isolation. The sun does not rise at all. Hunger, thirst and sleep become a way of telling the time. She crawls over the snow in the dark, harnessed to a makeshift sledge carrying ice from the glacier to melt and use as drinking water. She digs through a snowdrift and excavates the frozen remnants of a long-dead seal to cook on the broken old stove.

In case the reader should start to doubt that this can really be a true story, there are two black-and-white photographs at the back of the book: one of Christiane and her husband in front of the hut (which is if anything more rudimentary than she makes it sound) and the other of a huge snowdrift where the hut should be, and the two of them standing looking at it.

In Tromsø, they know how to use light to create spectacle. On the harbourfront buildings, coloured lamps throw their wobbling reflections across the black water. In the Arctic cathedral, the vast sheet of stained glass seems to store light in its brightly coloured cells like energy in a giant battery. Every shop window is strung with fairy lights, though in England they were taken down and packed in boxes at the back of the storeroom a fortnight ago. And the part of the town at the foot of the Fjellheisen is like a fairytale place, all gingerbread houses with twinkling windows, and lights in the trees dusting the snowy lawns with glitter.

It's not the Tromsø I imagined from reading the novels of Cora Sandel. Writing at the turn of the twentieth century, she evokes a city whose principal dimension is darkness, and a way of life structured and constrained by it. It not only determines everyday practicalities, but also shapes thought and feeling. Winter seems interminable. Streetlamps are a recent innovation, and limited to just a few streets in the centre of town. 'The arc lamps hung in the falling snow like pale moons', she writes. 'But a warm, red glow fell on the snow from windows and shops, making it bluer round the patches of light.' The face of the church clock also resembles a moon, and when it strikes to signal the start of day 'weak little lights took shape out in the darkness and burned dully, lost in its infinity, scattered and lonely'.

This is from her novel *Alberta and Jacob*, the first in the Alberta trilogy which draws heavily on Sandel's own life. It was another accidental discovery, picked up in a charity bookstall in Taunton railway station. (My reading life, like my writing life, is built on these random finds; as a result, it's an idiosyncratic structure, a kind of tall but rickety tower whose foundations are rainy afternoons, delayed trains and procrastination.)

Like *A Woman in the Polar Night, Alberta and Jacob* is a book saturated with darkness. But whereas Ritter's darkness is vast and wild, Sandel's has an opaque and oppressive quality that makes everything murky. People, landscapes, streets, boats, domestic objects . . . all are obscure and inscrutable. Alberta feels her way around indoors and out, bumping into things that loom up suddenly and frighten her. The winter dark has substance and texture; it fills up the inside of the house. A thick layer of ice on the windows blocks out what little light there is outside, and domestic chores like dusting require an intimate knowledge of the layout of the furniture and ornaments in rooms where 'objects rose to the surface like skerries in the ocean'. Human interaction, too, is veiled in darkness: meaning, feeling, motive are often indiscernible. Alberta's interior world is the whole world of the book, but her psychological and emotional life is shrouded and dimly lit. You have to strain your eyes to see what's there, driving her actions and especially her inactions.

Sandel was in her teenage years when electric arc lamps arrived in Norway in the 1890s. In her book they are still a novelty. Alberta takes regular winter walks with her father, but together they do their best to avoid having to interact with friends and acquaintances promenading in the well-lit centre of the city. They circumvent Fjord Street with its eight arc lamps, and walk along Rivermouth to the point 'where the last of the street lights ended and the blackness of the winter night began'. In these unlit streets, under the protection of darkness, they find the anonymity they both prefer. They encounter few of their fellow citizens in these obscure parts of town where no one goes except 'the energetic, the lonely, and those in love, people who deviated from the normal in one way or another'.

The Tromsø of *Alberta and Jacob* is hard to reconcile with the city today, with its brightly illuminated streets and colours shimmering on water. It takes an effort of imagination to see it as a place of deep winter darkness. What would Alberta, or Sandel herself, have made of all this gaudy radiance, not to mention the winter tourists, the film and music festivals, the bars and clubs spilling their brilliance onto the snowy pavements? I suspect both of them would love it. Sandel herself escaped to Paris as soon as she could get away. Her alter ego Alberta yearns constantly to be elsewhere. To her, the northern winter is gruelling and paralysing; it drains her of energy and potential.

But ambivalence ripples under the surface of the story. When relatives from the south come to visit, and express surprise that any kind of civilised life is possible here, Alberta feels scorn as well as envy. She dreams of escape, but does not take the chance when it comes. Convention is stifling yet safe. Darkness can be seductive as well as stultifying. She is drawn to those unlit streets at the edge of town, longs to escape the crowd and go alone and unnoticed there, in spite of dangers real and imagined. She is simultaneously repelled and fascinated by the mysterious hidden places frequented by her brother and his disreputable friends under cover of night. She feels the pull of those places, and of the different kind of life that is lived there, out of the glare of the arc lamps and free from moral scrutiny.

I wonder what has happened to those dark edgy places. Perhaps all this light has cleaned them away like disinfectant. And how far would I have to walk along Rivermouth before I came to that border point on the edge of the dark?

It isn't until we are on the island that I approach that border. We set off from the cabin and slither along the hard-packed snow, which gleams blackly under the streetlamps. Our walking boots are not adequate for these conditions; after a while I develop a style somewhere between walking and ice-skating, which would undoubtedly amuse the locals if there were any about, but we've only seen two people since we arrived and drove away from the quay, both making elegant progress using a kind of Zimmer frame fitted with skis.

There is just one road on the island; it runs along the shoreline for a mile or two before petering out at a rocky headland. Making our laborious progress we pass a scatter of houses, some with lit windows, a red-and-white chapel, and a small fish processing factory. There are lights at the factory, and the hum of a generator, though we see no one.

We are heading for the place where the streetlamps end and the blackness of the winter night begins. Our host Nora has told us that the last couple who stayed in the cabin set off to walk in the same direction, but turned back. They got close to the border point, and halted under the last-but-one streetlamp, terror-struck, unable to go any further. They were from São Paulo, she said, and had never encountered darkness like it. I just couldn't move, one of them told her, my legs said no.

We slide and stumble under that last-but-one street-lamp, and on to the last. For a few yards we follow the road in rapidly dimming light, before it curves to the left and the light is abruptly shut off by the sharp rise of the hill. It's as if someone has cut the power. On this moonless night, it's like walking off the edge of the world.

To begin with we can see absolutely nothing. It's so dark! I keep exclaiming. It's so *dark*! I'm expecting my night vision to kick in, but there's nothing. My pupils will have widened to let in more light, but the photoreceptors at the back of my eyes have been completely overwhelmed by the change in conditions and are not working at all. I feel my body faltering, its sense of space and balance thrown. Give it a minute. We stand and steady ourselves, each gripping the other's arm and breathing the splintery cold.

It takes time for the cone cells to adapt and some vision to return. It's like the line in the Emily Dickinson poem when the neighbour says goodbye and takes away the lamp, and 'A Moment – We uncertain step / For newness of the night'. This is a really long moment, though. I keep trying to blink away the dark, as if it has got into my eyes like smoke. After a minute or two I can just make out my gloved hand on my husband's arm, but I have to call on imagination to work out where I am and what surrounds me. It reminds me of pressing my eye to the keyhole of a locked church on holiday once, and visualising what I couldn't see: box pews, bell-ropes, famous rood screen.

The rod cells in the eye take longer to adapt, but after a few more minutes they catch up and take over. We shuffle on, con-centrating hard to stay on the road rather than plunging off the edge into the fjord. Our progress is so slow and unsteady, we must look very old, very drunk or both. Every now and then one of us exclaims over some ordinary feature as it materialises out of the dark – a stone cairn, a wooden post, a piece of rope.

It takes twenty minutes or so to reach optimum night vision. Now I can see details: tyre tracks on the road ahead, huge mounds of piled snow on the verge, even a boat on the rocky shore to my left. I have no idea what colour it is, though; rods are very sensitive to light, but no good at discriminating between colours. One set of powers is exchanged for another; it's like wearing a pair of special goggles that change my whole way of seeing. Now I'm giddy with the realisation that the dark is not as dark as I thought. I skate on over the hard-packed snow, smooth as a polished table. If it wasn't for the rocks looming up where the road ends, we could walk and walk.

Our jolly in Norway is a far cry from Christiane's experience on Svalbard. We have heating and lighting and even Wi-Fi in the cabin, and a small rather dismal shop where we can buy tinned and frozen food. There's a ferry to the mainland twice a day.

Grey Hook, where she spent that winter ninety years ago, is still one of the most remote places in the world, nearly two hundred rugged kilometres from the airport at Longyearbyen, but ships put in to nearby Woodfjorden in summer, and the hut itself attracts a particularly intrepid type of tourist. The experience she had just wouldn't be possible now. Svalbard is not as it was in her day; it is warming six times faster than the global average, and its sea ice is melting. The grief I feel over this is a form of solastalgia, because although I've never been there in person I have a dynamic sense of the place, gained from reading her book. I can only imagine how shattering it would be for her, and for the old hunters who made their homes and lives there, to see it so changed.

The book was originally published in German, under the title *Eine Frau erlebt die Polarnacht*. A literal translation of the word *'erlebt'* is 'experiences', which only increases the load of unspoken assumption. *A woman experiences the polar night.* Wherever this phenomenon of polar night might be found, and whatever it's like, it is by implication not commonly experienced by women; its conditions are in all likelihood hostile and challenging; adventuring there is thus the province of men, and its risks, thrills and hardships belong to them alone. No one, the title suggests, would expect to find a woman mixed up in all that. In fact, there were other women mixed up in it, including the legendary Wanny Woldstad, who arrived in 1932 and was said to be the first female hunter in the archipelago. Earlier still there was the unfortunate Ellen Dorthe Nøis, who in 1922 gave birth alone in an isolated hut, days after her husband had set off on skis to get help; there was a storm, he was gone for a fortnight, and she never fully recovered from the ordeal.

If I'm honest, I wouldn't volunteer for Christiane's diet of seal and ptarmigan eggs (said to taste 'of seaweed and mud'), the dread of being alone in a three-day storm without any matches, or the prospect of having to shoot a polar bear. The magic ingredient, the thing I envy, is the darkness. Sometimes she feels it annihilating everything, while at other times the sky affords some illumination. On one occasion there is moonlight so intense she feels she's dissolving in it, and on another the sky is pierced by northern lights 'like gleaming rods of glass'. When the sun rises, very briefly, for the first time on 25 February, the pale new light reveals the bleakness of the landscape, and she longs for the return of darkness, 'for the deep unknowing obscurity with its shining dreams'.

Here on our small island, the darkness induces a dreamy feeling in me too. In the middle of the morning there's a red flush on the horizon, followed by a couple of hours of exquisite blue-green light on the fjord and the hills. But it's not long before we slide back into the dark, into that feeling of spaciousness and possibility. It's as if time itself were suspended. Not being able to see my surroundings leaves more capacity for other things: thinking, feeling, imagining.

I've been thinking a lot about metaphor. The making of metaphor is one of the most fundamental forms of human creativity. Its intuitive leaps across space remake the world by subverting expectations and shifting perception of familiar things; it frees us from the limitations of the same few well-worn patterns of thought. I hoped that coming here and meeting darkness on its own terms would help me answer my own questions about its true meaning and character. Could I call on the visionary power of metaphor to change my perspective on darkness?

The metaphor I had in mind to begin with was *darkness as adventure*. I was thinking of a woman alone in the polar night, and a walk beyond the last streetlamp. Then I thought of *darkness as revelation*, which spoke of the riverbank haunted by strange lights, or the cave where the workings of time are fathomed.

But now, in this loose and dreamy state, I remember *darkness as refuge*, and the permissive space where we can withdraw from everyday demands and turn inwards. I think of the dark of the bedroom, the tent and the stock cupboard, those sanctuaries where intimate truths were spoken and the broken self pieced back together.

Writing about darkness is like moving around in it, I realise. Just as Alberta felt her way around dark rooms during the long northern winter, bumping into the furniture as it rose up out of the gloom, so I am feeling my way around my subject, bumping into stories, tripping over memories, using not only my eyes but my whole self to make out the shapes of things.

I wasn't sure whether it would be a good time to be somewhere so isolated. I am chairing the judging panel for a major poetry prize, which will be announced in London next weekend, two days after we're due to return. I should make it to the judges' meeting and the awards ceremony, as long as the ferry is running. (Heavy snow is normal here, and the infrastructure is built to cope with it, but there must be days when the weather is extreme even by local standards.)

While I'm here I have agreed to appear on *Front Row* on Radio 4 to talk about the prize. I've done interviews 'down the line' before, but it seems a stretch to imagine it working here. I will be sitting at this small table with my laptop and headphones, in a cabin with snow piled against the door, on the icy shore of the fjord, many miles from the mainland, and steeped in deep winter darkness. Will I be able to put myself back into that frame of mind, and to speak coherently about the books, the shortlist, the judging process? How real will any of it feel out here? And will the internet connection work?

A few minutes before the call comes through, an email pings into my inbox from a journalist on one of our national newspapers. They are running a story about the prize, drawing attention to what they call 'conflicts of interest'. All the judges are biased, they maintain – by personal connection, professional affiliation and imaginary quid pro quo – and as a result we have failed to shortlist the right books. As chair, I am especially culpable. The sneering implication is that I should not be in the role and cannot be trusted to carry it out with integrity. Do I deny it? Reply by ten the next morning or else. The wording is so aggressive it reads like a piece of hate mail.

When it's published a few days later, the story will turn out to be spiteful, ill-informed and too shallow to matter. But at this moment, already tensed for the interview, I feel a surge of anxiety. Snow reels against the black window. My mouth is dry. How easily I can be reached, in spite of the remoteness of the place. It's like being back in the worst of my schooldays, in that concrete playground where there were no shadows to hide in. Then the mood is broken by a crackle of static, and the producer's voice saying, 'Are you ready, Jean?', and I think, *Yes, fuck it, let them write what they like*, and I put on my headphones.

Ocean

My friend Joar at the Arctic University of Norway in Tromsø tells me that the long polar winter changes the way people live and work. You can come to work later and work a shorter day, and you're not expected to be as productive, though you have to make up for it in the summer months. Joar is one of the world's leading thinkers on human happiness, and we're chatting about the good life and what role darkness might play in it. By a real stroke of luck, Kari, an author and social psychologist who did her doctoral work with him a few years ago, is back in town when I'm here. I follow the two of them into the network of tunnels connecting the campus buildings, so students and staff can get around without having to keep going outdoors into the snow and ice. Kari is talking about her research on the 'winter mindset' that can make the difference between enjoying this time of year and enduring it. She started out with the common preconception that seasonal depression is more widespread among people living at northern latitude, but the studies she read suggested that this is not actually true. It seems that in countries like Norway, where the winters are long, cold and dark, societal expectations are different and people are able to adapt their mindset along with their way of life. Yes, says Joar, it's a matter of place. We don't want snow in the house, but we like to see it in the hills and woods. Personally, he enjoys this time of year; he can skate on the lake and ski to work. But then, he says, Norwegians know how to winter. He was surprised when he came to England to see people walking around in January without their coats on.

It's natural to need more sleep in the darker months. Until recently our working patterns were determined by the availability of daylight, and once it got dark opportunities to continue work were limited. People slept longer in the winter and worked longer in the summer; it was part of what gave the year its shape and kept them in tune with the rhythm of the seasons. Joar says he thinks this seasonal alignment is changing even here in the Arctic. It's possible not only to eat the same food all year round, but also – thanks to the ubiquity of electric lighting – to carry on working at a single unvarying pace regardless of seasonal rhythms. When a denial of the seasons sets in, our sense of who we are as human beings starts to degrade; we begin to see ourselves less as animals and more as machines.

In the Faculty of Biosciences, Fisheries and Economics, researchers are building an understanding of how the extreme Arctic conditions affect life on land and in the ocean, where long bright summers and long dark winters bring challenges that require all living things to adapt. Light is essential to plant growth, and during the polar night, when levels are dramatically reduced, photosynthesis slows and stops. For animals that rely on grazing and browsing, food availability falls away. In response, many creatures enter a period of winter dormancy. Indeed, it used to be assumed that biological activity across the entire range of Arctic species was minimal during this time of year. But it turns out that not all living things are affected in the same way. Some organisms respond to light levels too low for humans to detect. Some go on feeding and reproducing, even through the very darkest time of year. Polar cod continue to hunt for prey, though how they find it in such dim conditions is still not fully understood. Even more impressively, cod has evolved to become more reproductive during the winter months, shifting its spawning cycle so that the larvae will hatch and feed for the first time just as the availability of prey is at its maximum. Over thousands of years this species of fish has adapted not just to survive the dark but to depend on it.

Arctic seabirds, too, are active through the polar night. Seabirds are thought to be visual hunters, relying on their eyesight to locate their prey as they dive deep underwater. However, a study of great cormorants recorded them diving for fish every single day of the year, summer and winter. They made no adjustment to the rhythm of their hunting to fit in with the varying length of day. These birds were diving through dark air into dark water, and still, somehow, locating their prey with pinpoint accuracy.

Water absorbs light strongly, like a sponge absorbing spilt ink, and suspended particles in the water scatter what little there is. At a depth of one thousand metres, however clear the water, almost no daylight penetrates. In order to survive in these conditions, aquatic animals have developed varied and specialised eyes. The deeper the habitat, the larger the eyes become; the eyes of the giant deep-sea squid measure thirty centimetres in diameter. These enormous eyes have evolved not to take in daylight, but to detect clouds of bioluminescent plankton, because when plankton are disturbed it means sperm whale are nearby. For ocean creatures that are prey to the sperm whale, the capacity to see this early warning sign makes the difference between being eaten or getting away.

*

When Christiane Ritter talks of darkness as death, she is not reaching for the usual well-worn euphemism. She is recounting what she sees all around her on Svalbard. She repeatedly uses the word 'dead' to describe the landscape, noting again and again the bareness of the rocks, the absence of trees and vegetation. In the depths of the polar night, sightings of animals and birds are rare and therefore momentous. Winter storms rage for days on end, making way for periods of eerie moonlit stillness. On one occasion she observes a halo around the moon, which blazes like 'white fire' in the intense darkness. 'Against that tremendous burning circle of light,' she writes, 'the earthly scene is dead and shadowy, an extinct orb.'

But she lives there long enough to learn that this frozen landscape is not really dead. Even at the darkest time of the year, she observes eider duck, ptarmigan, seal, fox and polar bear. After a while the tables are turned: she begins to associate Svalbard with life and Europe with death. She forms an attachment to an Arctic fox which she names Mikkl, and is determined to save his life. Warning him about the fox trappers and the fate that awaits him if he is caught and his fur exported to be sold in the city, she says: 'There they will give you glittering eyes made of glass, and then you will hang in one of the thousand glittering shops in one of the thousand glittering streets, together with thousands of other glittering dead things. Do you know, Mikkl, there's so much artificial glitter there that the people no longer know anything about light, about its coming and going, and about the magic of twilight.' She has come to recognise the interplay of light and dark as the very essence of the place, the source and expression of its life.

At the end of her adventure, when she is on her way home,

she is able to look back on this lifechanging experience and articulate what she has learnt: 'You must have gazed on the deadness of all things to grasp their livingness.'

In 1902 Fritjof Nansen, one of the earliest explorers of the Arctic, gazed on the deadness and arrived at his own view. He and his team made observations of the areas covered by ice, and concluded that they were 'unproductive'. With the technology available at the time, they were not able to detect signs of life in or under the ice. At the same time, however, there was evidence of human settlement in the high Arctic. How could people have subsisted in regions that looked like biological deserts?

It would be seventy years before instruments were developed that made it possible to observe life forms not only beneath but also within the Arctic sea ice. Now we have images that show these vast sheets of ice blooming with bioluminescent colour from the numerous species of marine invertebrates that live there. Under the ice, plankton, crustaceans and fish survive the extreme conditions in their own individual ways. Some make a seasonal migration downwards into deeper water, where they overwinter in a state of suspended animation. Others are more active, migrating daily in pursuit of food. This regular movement from shallow to deep water and back again is regulated by the circadian clock, that miraculous piece of engineering that enables living things to anticipate the future. These marine organisms – apparently simple, some microscopically small – have the capacity to predict when dusk will fall with such accuracy that they are able to set off from the depths to travel upwards and arrive in shallower waters at exactly the right time.

During the course of evolution their internal clocks,

like ours, have developed in response to the most stable and predictable of routines: the regular rotation and orbit of the earth. Light and dark are the most important of the zeitgebers or 'time givers' that keep the clock synchronised, entraining it to a twenty-four-hour rhythm. In polar regions, where the cycle of day and night is absent for several months of the year, the animals that live there have adapted to survive without the usual zeitgebers. Like Salvador Dali's clocks, draped in distorted and molten forms over the branches of trees, the circadian clocks of reindeer on Svalbard lose their definition in summer, when it doesn't get dark at all. This allows them to forage continuously and build up the fat reserves they rely on throughout the year; there are only a few weeks of plant growth, so there's no time to lose. At such times these animals are capable of transforming their behaviour in ways that would stress a human being to death. They go without rest during this brief window of opportunity, and their hearts run more than twice as fast as usual. I imagine them pushing on and on, in spite of the time, like a writer with a deadline working frantically night and day to get the chapter finished.

My friend David, who has sailed all his life using the disappearing art of celestial navigation, speaks passionately about the pristine quality of darkness he experiences in mid ocean. Once you're a hundred metres from the nearest coast, he says, no light pollution from land will trouble you. Occasionally there's the light of a ship or an oil rig, but mostly it's just you and the dark. On watch in the middle of the night, when there is little else to do, you can observe the movement of the constellations as the planet rolls through space. You live in the moment, he says, at the mercy of the elements, entering into a different relationship with time, and with all the life around you.

To think of places where darkness is at its most impenetrable – the benthic zone, the deep cave – is to be reminded that in spite of our human culture, science and medicine, all the trappings of civilisation, there are limits to our knowledge and mastery. Technological advances make it possible for explorers to go further and deeper, and to shine their lights into spaces never seen before – the moon's craters, the Mariana Trench, the Veryovkina Cave in Abkhazia – but one question gives rise to another. A recurring feature of these expeditions is the discovery that darkness in such places is less complete than we thought. Numerous species of fish, jelly and other deep-sea creatures use bioluminescence to lure prey, confuse predators or attract mates; and researchers at NASA recently announced that there is approximately twice as much light in the universe than previously detected, and most of it cannot be accounted for. 'Perhaps there are unrecognised galaxies out there,' said one of the lead scientists, 'or some other source of light that we don't yet know.'

The Arctic is full of life, even in winter. Under thick sea ice, critically endangered bowhead whales glide through the dark water, singing all through the polar night. They wouldn't be audible to David as he stands on deck and watches the stars, nor to Christiane walking the moonlit shore. They are not singing for us. But when acoustic recorders were placed in the ocean off the coast of Svalbard, they picked up calls and songs occurring almost every hour between early November and late April. Winter is their peak breeding season, and the singing males like to perform in the coldest and darkest waters, preferably under a dense canopy of ice, perhaps because it creates more favourable acoustics. Anyone who has ever sung in the bath will have noticed how good they sound in that tiled and echoing space, though it's doubtful at best whether this makes any difference in attracting a mate. For the male bowhead whale, though, it's a brilliantly successful strategy. In the intensity and complexity of their singing, and the sheer diversity of their repertoire, they are said to be more like song-birds than whales.

These recordings are a world away from the humpback whale song we're all familiar with: that sonorous, dignified music whose very restraint gives it a timeless and haunting quality. These are more freeform, more avant-garde. Listening through headphones, I feel I'm tuning in to another realm of darkness altogether, one where the usual assumptions are turned on their heads. This is a darkness whose meaning is *life*. I want to laugh out loud with the sheer exuberance of it all. One piece sounds like tyres on a wet road when you take a corner too fast, the next like the high-pitched, expressive whistling of the Clangers, and the next like a teenager riffing experimentally on an electric guitar. There's a particularly lovely number in which a gruff bark alternates with a high plaintive sigh. I close my eyes and revel in this strange and lusty music as it echoes down the vast arcades of the ocean, far beneath the starlit roof of ice.

Hut

To the east, a fiery red glow presaging sunrise. But the sun doesn't rise – instead it very gradually dissipates. The fjord is aquamarine, with little folds or curls of light and shadow travelling across it. To the west, the snowy hillsides are tinted apricot pink, and the sky above them is green. The lamps are twinkling along the road, but there is no sign of anyone. Then a single car comes crackling over the ice, crawling east towards the *ferjekai*.

A space has opened up in the darkness. It will close again soon, when the sun drops further from the horizon and the blue hour softens into black. Then we are going to venture out for a real walk, this time away from the road and the lights, uphill into the uninhabited centre of the island to a hut where we can light a fire. Nora has lent us snowshoes, and we have practised in front of the cabin, waddling flat-footed like ducks.

In one of the books left by a previous visitor, I read about a linguist called Ole Henrik Magga who has established that the languages spoken by the Sami peoples of northern Scandinavia and Russia include more than a hundred and eighty different words for snow. Previous claims of this kind have been the subject of controversy, but even a brief visit to this part of Norway in winter makes Magga's calculation feel credible. The gleaming lines of snow compressed by the snowmobile are a very different phenomenon to the mountainous heaps of dirty snow shovelled at the edge of the car park, or the thick snow glittering under the lamp outside our cabin, dry and soft under its crisp surface. If you work, play, travel and live for many months of the year with all these different kinds of snow, each possessing its own particular character, you need a diverse vocabulary in which to talk about them.

I enjoy the walk up from the fjord, first between hedges and a small field with goalposts sunk deep in the snow. In the low light, snow obliterates the contours of kerb and verge and pothole, erases the distinctions between lane and track, smooths everything together in an illusion of seamlessness, like a dream. Then the residual light of the road dissolves and we are really in the dark. Without the distraction of sight, my hearing is sharp as a blade. The snowshoes make different sounds as I go over different surfaces. They clatter noisily on the impacted snow, which is as hard as a road where the snowmobile tracks are. Then they yawn and groan on the deeper snow, as if all the empty spaces in its loose structure are alive with voices speaking back.

Odd echoes skitter around the valley. Distances are hard to judge; each sound stands out in detail, pulling everything unnaturally close. The sky is moonless, and when I stand looking up I lose my sense of space for a moment as the white-hot stars press urgently towards me.

The two of us have gained our night vision and become more confident on the snowshoes. A comfortable distance has opened up between us as we each find our own way over the same terrain. Then suddenly there's an unfamiliar noise, and instantly I snap back into alertness. Unmistakably an animal noise. It was quite loud; surely close? Or maybe further than I think. Maybe somewhere among the sparse trees I can make out to my left. I call my husband's name. Where is he? We've become separated somehow. Terror flashes through me as I fumble with my inadequate torch. Pointing it at the ground ahead of me I see I'm lost – I've wandered far from the snow-mobile tracks, and there are sharp rocks cropping out of the snow, and dips and hollows that might open up into crevasses. I blunder away, stumbling, tripping over my snow-shoes, shouting that we need to get to the hut *now*.

My night vision is ruined and all I can see is what appears inside the disc of torchlight. It seems ages before I find the faint silvery line left by the snowmobile. I follow it blindly until I see the hut, trembling like a mirage up ahead. I shove open the heavy door, and the wind roars in off the mountainside behind me. Shove it shut, have to push hard against the snow. Point my small torch nervously, track it round the space.

Just then there's a loud scuffle outside, and the door is shoved open again and he bundles in, the wind and snow chasing in after him. What was it, he wants to know, but I can supply no answer. I steady my breathing and wait for my heart to slow to normal. We're safe now. Whatever it was, it can't get us here.

In the feeble torchlight I can see a fireplace made of large stones in the centre of the hut. And here's a heap of firewood. We make firelighters by tearing pages from my notebook and tying them into double knots, as I learnt to do on a camping trip when I was a kid: it was a miserable, wet weekend in a field near Tamworth, but today I feel a stab of gratitude for this useful basic technique. The firewood is furred with ice, but it must be dry underneath because it catches readily to life.

Now we can see where we are. The hut is a conical structure made of birch logs in the Sami style, with a low wooden bench running along three sides of it. Smoke is drawn up like grey wool on a spindle, and escapes through a hole in the centre of the roof. Snow has drifted in under the door and lies frozen hard and glossy in the firelight.

Now by the fire I can tell myself it was irrational. He's too kind to make a joke of it, but I do it myself. What do you think is out there, Jean – wolf, polar bear, yeti? It was nothing. Probably some kind of owl. In all honesty I know perfectly well it wasn't really bears or wolves I was thinking of but something outside the sphere of ordinary experience, something supernatural . . . but I'm too embarrassed to admit it. Let's just stay here for a while, give it time to go away.

It was the darkness – it had me under a spell. The stars made me dizzy. There I was, out in the open, with the power of sight stripped away, and a hair-trigger alertness to sound, and when I heard one I just freaked out. Reflecting on it now, it seems to me that we need a hundred and eighty words for darkness, and one of them would be exactly right to describe this particular deranging variety.

But now all I feel is the contrast between the firelit interior of the hut, and the huge overwhelming darkness outside it. As Joar says, it's a matter of place. How safe I am in this simple shelter, how extravagantly grateful to whoever left the heap of firewood. My heart aches. Smoke stings my eyes. Ice melts and pools between the stones.

Heaven

In this church, there are several pieces of devotional sculpture featuring a sunburst – a ring of splintered silver or golden rays at the head of a saint or the skirt of a cloud. Of all the ways artists have attempted to represent light, this must be one of the least sophisticated; it's suggestive of a child's drawing – it doesn't actually look like light.

It's misleading to call it a representation at all; it's more a kind of gesture, placed there to remind the viewer of certain qualities of light: its suddenness, the way it flashes and splits, grows angles and points. This is not an effort to imitate the inimitable nature of light, its mutability, its capacity to make and remake the world, the way it shines through leaves, moves over water, draws shape and line, lends lustre to skin. That impossible endeavour is left to other kinds of art in other places. For these sacred walls, the church fathers commissioned work that would tell a story and to stir the hearts of the congregation with awe.

These spars of painted wood remind us of the role of light in making visible what really matters; in seizing attention, focusing, clarifying. The figures are powerfully lit from above and behind, which in reality would throw them into silhouette and make detail hard to see. This is a recurring motif in scripture: sudden, overwhelming light which dazzles and temporarily incapacitates, forgoing complexity in favour of the single blinding truth. The light from heaven that strikes Saul on the road to Damascus; the angelic light that terrifies the shepherds near Bethlehem; the supernatural light that floods the cell where the disciple Peter is being held.

The eye can only adjust to so much light. In extreme illumination, some things are made decisively, unambiguously clear, but others are lost to invisibility. Sometimes that clarity and simplicity are exactly what's required: moments of decision-making, moments when we need to recognise a truth or hear a call to action. But too much light scours surfaces, erasing detail, leaching colour, imposing a regime of brilliant uniformity. It chases mysteries out into the open and irons ambiguities flat. Light like that leaves nothing to the imagination. It can do away with both faith and doubt at a single stroke.

Then shadow brings relief, allowing once more for subtlety, revealing depth and texture, opening a space in which the religious mystic reaches through doubt and prayer for the divine.

There are times for the single blinding truth, and times for nuance and questioning. Light and dark take turns to speak and listen, temper and augment one another. What do they know or care of the human wish to work by night, to travel in safety or to claim respite from the knowledge of death? They are neither opposites nor enemies; not two distinct states but one endless conversation, with its own mysterious lexicon of courtesies and raptures.

small fiery exclamations among the steady plot of the stars

You had climbed a hill and lain down on the grass together to look at the sky as it darkened. You were surprised how long it took; the greens and browns and greys of the field and trees and scattered stones of the ruined sheepfold had given up their colour, but the sky still brimmed with light. Then the first stars – however hard you tried to watch them appear it was impossible, they were just *not there* and then *there.*

You were twenty-two, it was August, the air was warm, and the darkness made things feel limitless. Above you, space opened up and you could let go and fall into it if you wanted. Your body was only lightly connected to the ground, and gravity would not stop you if you wanted to go. You didn't believe in fate. It wasn't that your names were written in the stars or anything like that. It was a moment, and you could choose to fall.

You must have slept, because then you heard him say your name and you opened your eyes and there were shooting stars. Just a few to begin with, caught in the peripheral vision, and then more, and more: hundreds of small fiery exclamations among the steady plot of the stars. They chased from the crown of the sky towards imaginary points on the horizon, and burned out as they fell.

It was exactly like a dream. But it can't have been a dream, because you noticed as you walked down the hill at dawn that you made two sets of footprints in the dew.

Wassail

If it's darkness
we're having, let it be extravagant.

— Jane Kenyon, 'Taking Down the Tree'

Pond

When they came to dredge the old pond and take down the willows that had rooted there, they found eels.

The pond is shallow and silty, fed by water running down the hillside from springs that rise in the fields above. A bank a few metres wide separates it from the river. To make their way into the pond, they must have travelled over land, at least for those few metres. I was astonished to learn that they could do this. Fish moving over stones and through undergrowth between one body of water and another ... it was hard to imagine. I could only visualise them slithering like snakes through the long grass; they are snakelike in shape, and they move sinuously through water, so it seemed just about possible.

Mac, who was in charge of the pond work, was only mildly surprised. Eels are often discovered in ponds, he told me. There was a rumour of eight-foot specimens in London's Hyde Park. It was probably much more common than we thought, he said, because their preference is for the murky depths where they remain hidden unless disturbed by restoration work like this. Our pond had been neglected for many years, and they had been able to thrive without disturbance until Mac and his team came along.

I sat by the pond at owl-light and watched, hoping to glimpse them. It was calm and introspective. Bats worked above the water, snipping insects out of the air. Strange how sleek their movements are, I thought, and how different they look when you see them roosting: small, leathery and angular. At roost they resemble the odd implements and receptacles we found on shelves and hanging from nails when we moved here – tin jugs, twists of rope, an end of thick stuff that cracked and disintegrated when we tried to unroll it – but in their element they are slick and exact. Meanwhile no sign of the eels, no break in that darkening surface.

Nevertheless, there's every reason to think they are still there, living peacefully in the silt and shadow. I wish I could spot one as it slinks through the water, or better still over the grassy bank, but their overland journeys are almost certainly made at night. And this short terrestrial venture is by no means the most extraordinary part of their travels. Each individual eel in our pond started life in the Sargasso Sea, where the egg hatched into a transparent larva just a few millimetres long. It entered the Gulf Stream and was carried around the North Atlantic gyre. By the time it reached the continental shelf of Europe, it had grown to ten times its original size and metamorphosed into a glass eel. It found its way into the shallow coastal waters off Dorset, and subsequently into the mouth of the River Axe, where it changed again, becoming first an elver and then an adult or yellow eel. It swam upriver, and just south of Axminster it turned off into the River Yarty, following it for almost all of its twenty-six kilometres through Westwater and Beckford, under the A30 at Yarcombe and up into the hills until something told it the place was right and it hauled itself out and over the bank and into the pond that would be its home for twenty years. One night, when it has reached sexual maturity, it will slither out and set off to make the same journey in reverse, transforming once more into a silver eel as it covers the five thousand kilometres to the Sargasso where the next generation will be spawned.

How this migration is achieved is still somewhat mysterious, though it seems that eels are able to detect the earth's magnetic field and use it to navigate their complex journeys over vast distances. Geomagnetic navigation is a rapidly evolving field of research, but grains of magnetite are known to exist in some birds, insects and fish – in the beak of the pigeon, for instance, and the abdomen of the honey bee – and to play a role in their ability to orientate themselves in space. It's thought that this internal compass makes possible feats of navigation beyond ordinary human comprehension.

Eel journeys take place at night, and depend on the accurate functioning of the biological clock, which is synchronised with the internal compass. The dark-light cycle that entrains the clock can be badly disrupted by light pollution, and this may be one of the reasons why the European eel, once very common, is now critically endangered.

It is often said that we are living in dark times, but the metaphor could not be more wrong: in fact the times we live in are perilously light.

Artificial light spreads further and faster than we imagine, not only in cities but in remote rural settlements and even out into the ocean, where oil rigs and shipping lanes break up the dark and fisheries blast the water with bright light to attract squid. We are putting life in jeopardy with our efforts to snuff out the dark, because every living thing needs it, from eels to beetles, from bats to fungi, from trees to people. Without darkness, we are lost.

Eels are a strange sort of fish, not only on account of their snaky appearance but also because they can survive out of water for several days, thanks to their thick skin and relatively low oxygen requirement. Their life cycle is convoluted, and their migratory journeys immense, yet it's the comparatively short distance they can travel over dry land that entrances me most. Perhaps 'dry land' is not quite the right term to use, since they seem to favour grass slicked with rain or dew. But there's a poem by Thomas Hardy called 'Night-Time in Mid-Fall', in which it seems they are prepared to venture over even less hospitable terrain, and 'Even cross, 'tis said, the turnpike-road'. With that disclaimer, "Tis said', the speaker implies that it may just be a tall tale. But then he adds a note of corroboration: '(Men's feet have felt their crawl, home-coming late)'. It's an unlikely detail, tellingly placed in parenthesis, that muddies the water still further. Either way, it's grisly enough to make me shudder as I sit by the pond and wait. While I'd love to see an eel, I definitely don't want to step on one.

Crevice

When the military helicopter comes over I'm on my hands and knees in the middle of the lane, face to the tarmac and backside in the air. I hear it roll towards me in the dark, and now it seems to slow and stall overhead, hanging so low I can feel the draught from its rotor blades. Light and noise rains down on me. I do a quick calculation in my head and decide it would be more embarrassing to look up than to stay where I am and pretend I haven't noticed, so I hold my ridiculous posture and imagine the young pilots in the cockpit craning their necks, pointing and laughing at this strange and undignified spectacle. What *is* that woman doing?

What I am doing is pressing a hand lens close to a tiny rift in the road surface, and a glow-worm positioned on the lip of the crevice. The pilots peer down at me, and I peer down at her, and at that moment it would be hard to say who is more mystified.

I found her in the lane, just outside our gate. She is a chip of bright green light that I could have mistaken for an LED come loose from a camera. When I hold the hand lens close, she is nothing but a blur of green to start with. Now, as the noise of the helicopter fades and vanishes, and I focus more carefully, I can make out the shape of her body, and the pattern of illumination on her abdomen: two crescent-shaped stripes and a row of three short dashes. Under magnification I can see that she is rocking slowly from side to side, a movement too slight for the naked eye. She isn't just passively emitting light, but *signalling*.

Since the beginning of human history, people have observed unexpected light glinting on the ground, sparkling in the air or shimmering in water. These were not will-o'-the-wisps or Jack-o'-lanterns but demonstrably living things with the capacity to generate luminance from within their own bodies. Aristotle alone described a hundred and eighty species that produced this 'cold light'. In different times and places a range of hypotheses were used to account for them. Because they were so unearthly in appearance they were often thought to have paranormal significance of one kind or another; fireflies feature in some of the earliest religious writings from China and India. Milky seas – large areas of the ocean which shine silvery blue – were particularly enigmatic. They were also known by the beautiful name 'mareel', which comes from the Old Norse for 'sea fire', and tales of them, brought back by mariners, must for centuries have seemed like the grog talking.

Milky seas sometimes cover hundreds of square miles, and are large and bright enough to be visible from orbiting satellites. But other instances of bioluminescence are much smaller and less photogenic. The glow-worm or *Lampyridae* is so modest that it often goes unnoticed by humans. I first saw them many years ago on a camping holiday in France – dozens of them low in a hedge and on the grass beneath, like fairy lights. I felt sure they also existed in Britain, though I'd never seen one. My initial research was confusing – some sources claimed the glow-worm was a type of firefly, others that it was a beetle. Either way, it definitely wasn't a worm. And although it wasn't rare in this country, most people had never seen one. My casual interest flared up into an infatuation.

They may not be rare, but spotting them is a challenge: you have to get your eye in. There is no single recipe for success when hunting for them, no particular species of plant that can be used as an indicator. They turn up in some improbable places. But there are some guiding principles. In order to thrive, they need a mosaic of open areas with no obstructions in the way to prevent the flying male from spotting the glowing female, and areas with a broken structure that provide access to an underground chamber for laying and hatching. They prefer poor soil and unimproved grassland. They are found in weedy areas, but not in deep undergrowth. They turn up in woodland clearings, but not under the canopy of trees. (Sara Teasdale's poem 'August Night' places one in 'a wood too deep for a single star to look through', but this sounds improbable. But then the poem is only superficially about the glow-worm, and it does contain a marvellously accurate image of it brightening 'as though it were breathed on and burning'.)

They particularly like crevices in stonework, or churned-up soil of the kind found on road verges, cemeteries, abandoned quarries and coast paths, or among the ballast on railway lines. Railway lines are especially favoured, and it's common for a population to remain in place when the line is abandoned and dismantled by heavy machinery, even when the ballast has been removed and the route converted to a footpath or cycle track. Despite all the upheaval, they go on with their quiet, unobtrusive lives as if nothing at all had changed.

I have learnt most of this from Robin Scagell, a legendary figure amongst glow-worm enthusiasts and the undisputed authority on these tiny and easily overlooked creatures. He was easy to find online, because a search took me straight to the website of the UK Glow Worm Survey, which he established over thirty years ago. When I first came across his name, I couldn't shake the feeling that I'd heard it before. It's an unusual name, one I had to learn to pronounce correctly. (My first attempt, he said, made him sound like a relative of Arthur Scargill.) It wasn't until he mentioned his interest in the night sky that I twigged: I owned a book of his, and it wasn't about glow-worms but astronomy. The book was *Philip's Complete Guide to Stargazing*; I bought it when we moved to the cottage and I wanted to understand more about what I could see in those startlingly clear night skies. When I retrieved the book from a shelf in the hall and read the blurb, I realised that he is in fact an eminent astronomer, and author of several books on the subject. It occurred to me that astronomy and glow-worms are a perfect pairing. To study both would make for a rich and rewarding life for someone who enjoys being out in the dark. Each has its season, so that there are always stars to look at, as long as the weather is right: in winter he finds them in the sky, and in summer he finds them on the ground.

Robin's own infatuation began when he saw glow-worms while working overseas, and on his return he set about investigating what populations existed in the UK. The Natural History Museum supplied him with a list of old records compiled in the 1970s, and he embarked on a project to match them up with present-day sightings, with the aim of finding out whether the same locations continued to support populations of glow-worms twenty years on. Some of those original records were vague and unspecific, such as '3 miles S. of Bath' and 'Longhope area, two sites c.1 mile apart'. It took time and patience to comb through the list. Then each site had to be visited during the summer months, the glowing season. It was obvious from the start that he couldn't do this alone; he would need a lot of help, and that meant stirring up public interest. He wrote articles and gave interviews, raising awareness and appealing for volunteers to go out after dark and look for glow-worms in their local area. In the decades since, he has continued to lead an extraordinary piece of citizen science: a nationwide movement to record glow-worm numbers and create a living map of where they are found.

The picture that has emerged is one of remarkable consistency. It turns out that colonies of glow-worms are settled communities: having established themselves in a particular place they will continue to live there, as long as the conditions don't deteriorate too much. They are creatures of habit, and don't spread easily from one place to another; they tend to stick to home ground, being observed there for many human generations. Attempts to seed a new population, for instance by taking individuals and putting them in a garden, are invariably doomed to failure.

Looking at the survey map, I see there are huge geographical variations. In some regions there are no sightings at all. It seems that when glow-worms colonised this country after the Ice Age, none made it into Lancashire, the Wash or East Riding. The more flooded parts of the Somerset Levels yield no results either. They are rare in cities, though a solitary sighting was recently made in Central London, causing a brief flurry of excitement until it was found to have been imported on a plant bought at a garden centre.

Even on the longest of summer days, in the most brilliant sunshine, when the valley rings with birdsong for eighteen hours or more, this place is slow to forget the night just gone, and early in anticipating the next. Morning sun takes time to reach the valley floor and burn off the dew. Many hours later, with the coming of dusk, catchflower unclenches its creamy white petals, and hedge bedstraw releases its scent in readiness for moths, and the bats that feed on them.

If you want to see glow-worms, there's no point in going out too early. You have to wait until colours ebb away and monochrome tones take their place. That was my cue tonight to set off and walk uphill to the plateau. On the short ascent I passed through successive layers of heat, of stillness and of darkness. There were quiet rustlings in the hedgerows, and waves of perfume from the honeysuckle and the wild rose that threaded their way among the hawthorn, elder, maple, ash, rowan . . . these hedgerows are so old and so rich you could lose time counting the different species they're made of.

I stopped to catch my breath at the top of the first steep climb. No sign of a glow-worm yet. I had seen them beside the path in the conifer plantation north of here, when I was out in search of nightjars, but never in our own valley. I was hoping tonight would be the night. In the distance a column of red lights twinkled on the radio mast, and I was reminded suddenly of standing on the beach at Southport and looking at the lights of the oil rig shivering over the water. I felt a pang of loss – not loss of the place itself, but of those times when the children were young and the four of us stood there on the sand together at dusk, before heading home in the tidal dark to light the fire and make tea.

They were always excited to be out at night. They would hang on to our hands and exclaim as they watched the white edges of the waves scudding towards us out of the dark, then run ahead to hide behind a wall and jump out on us. Looking back, I see that it was neither pure pleasure nor pure fright they felt, but a delicious intermingling of the two. On camping trips they loved to lie on a blanket and look at the stars, and to walk to the badger setts and sit very still till the cubs and their mother came out to play and hunt. In the stillness, the snap of a twig could make us all jump. Part of the thrill of the dark was feeling a bit scared.

How simple life was back then, I thought, though really I know that memory is deceptive, has a way of smoothing out the complexities. I remember when they were older, and going out with friends in the evening, how I would lie awake in the dark, waiting for the sound of the key turning downstairs, the door bumped shut or pushed softly to, shoes kicked off in the hallway. In the hinterland between dream and wakefulness, I would rehearse the same dire scenarios – late-night jaunt with learner driver at the wheel, midnight swim in rough seas, unknown drugs in scuzzy nightclub – none of it so very different from my own teenage escapades and narrow scrapes, which made it worse because it meant I could visualise it all in agonising detail. There was no way I could sleep till they came in. Then there was a feeling of rightness, everyone safely gathered under my roof and order restored.

An insect tickled my bare arm, and I was back in the lane, in the hospitable darkness of this midsummer night. The warm smell of the fields, an echo of birdsong. I wanted to take my time in this scented and luxurious dark, to dawdle home and make the most of it, because in only a few short hours the sun would be up again. Where are the metaphors that speak of this benevolence, I asked myself, this delight and sense of belonging in the dark? *Darkness as enchantment*, perhaps, the way it transfigures fields and woods, streets and parks and riverbanks. I walked on, brushed the silken net of a spiderweb, and felt bad about breaking it – they take such time and skill to construct. Pieces of discarded straw in the lane were incandescent, and tiny stones on the verge glinted like gems. At a fork in the road I took a left and climbed again, and there was the moon, resting on a field hedge: ruddy faced, squarish, fuzzy at the edges. Dear old moon, I thought, you often seem so cold and remote, but tonight you're like a friendly countrywoman greeting me as I pass. I smiled in return, and when I looked away I could see the fields gleaming as if touched by summer frost.

It's mid June when I'm kneeling arse-up in the lane: peak glow-worm time. Summer is their season, though dates vary from one year to the next depending on weather: they can tolerate a degree of wind and rain, but they really don't like the cold.

Successful breeding in any year depends entirely on the female's ability to display her light and the flying male's ability to see it from the sky. She will station herself in the same location, on the ground, or on a stem of grass close to the ground, night after night from the earliest viable date until she has attracted a mate. After mating she retreats below ground, lays her eggs and dies. When the larvae hatch they feed on snails, which they inject with a fluid that dissolves the internal structure and allows the contents to be sucked out like soup.

Their numbers seem to be declining, but it's hard to be certain because there are such dramatic fluctuations from one year to another. One summer night, hundreds of signalling females congregate unexpectedly on a hillside where the year before there were none at all. It's not easy to draw general conclusions based on these ups and downs in their fortunes. But given the collapse in insect populations more generally, it's likely that they are becoming more and more marginalised by pesticides, by changes in land use and especially by artificial light. Anecdotal evidence suggests this is true. Older people recall walking home through their villages at night guided by glow-worms along the edges of a dark road; then streetlamps were installed, and no one sees a glow-worm there anymore.

Perhaps the streetlamps trick the female into thinking she is in an open sky area, good for transmitting her signal. But then the artificial light drowns out her own and makes it impossible for the male to see as he flies overhead. Street-lamps, floodlights, spotlights, security lights – they are pro-liferating, and brighter than ever, and many are left on all through the night. Where the female's speck of green light would in the past have been visible from a distance, our human lights – many hundreds of times brighter – blot it out altogether, and mating fails. It doesn't take long for this onslaught of extra light to wipe out a colony.

High-volume light is not the only problem; the glow-worm's very particular characteristics make it vulnerable to smaller and more discreet sources too. In recent years it has become fashionable to place coloured lamps on graves in cemeteries, and male glow-worms have been observed visiting them in the belief that they're females signalling for a mate. I guess some people find it comforting to think of a light shining all night, marking the resting place of the loved one with battery-powered *lux perpetua*, but we make a more meaningful legacy by guarding our cemeteries as places of sanctuary for wildlife. Out of respect for the living as well as the dead, their natural dark should be sacrosanct.

The records show that 2020 was a bumper year. The sight-ings flooded in and the survey map bristled with pins. It might have been a consequence of the glorious weather we had that spring and early summer. Or it might say more about us than the glow-worms. Perhaps it had something to do with the Covid pandemic, the spring lockdown and restrictions on movement. For a time we were all confined to our local areas, unable to go far from home and follow our usual patterns of work and leisure. It could be that these constraints changed the way we looked at things. We may have sought out sources of joy and interest closer to home; perhaps we were more willing or able than usual to spend time outdoors, and more inclined to follow up by counting and recording our discoveries. The problem with counting tiny creatures like glow-worms is that it relies on people noticing them, and most of the time we simply don't. No one's going to spot one from a moving car as they dash between one place and another. We have to slow down.

Now my knees are starting to complain – it's time to tear myself away and leave the glow-worm to her signalling. Tomorrow I'll walk up to Top Stile with my phone, log on to the survey page and drop a pin on the map to mark my sighting. No problem identifying the exact place to drop it. How miraculous that she should turn up right here at my front gate! It's tempting to feel I have conjured her up, but she was probably here all along and I just wasn't looking. Either way, she has chosen this spot because it has the simple things she needs: stony ground, open sky and unspoilt darkness.

I once asked Robin how he felt about darkness, since he spends so much time in it. He told me that he loves twilight, for its colours and tones in both landscape and sky. He likes the dark of night too: it's immersive, he said, rather like swimming. But he had to learn to be comfortable in the dark. He recalled a night at Dawlish Warren, when he was fourteen or so and just becoming seriously interested in astronomy. He had left the lights of town behind to walk out alone onto the sandspit across the mouth of the estuary, a long stretch of shoreline, mudflat and sand dunes. He stopped dead, he said, suddenly afraid by the darkness, which was more extreme than any he had encountered before. He had to coax himself to go on.

To be afraid, and to coax ourselves on: this is how we learn to make peace with our fear. We can't live in the light all the time. And small adventures after dark can give us new ways of seeing that aren't available by day.

I was repeating this thought to myself as I started back down the hill tonight. Cloud had smothered the moon, and I was trying to keep faith with the small adventure and the new ways of seeing. No luck again tonight. Maybe there were no glow-worms in this valley after all. Maybe I needed to accept that this was one of the empty spaces on the map. You can't just wish things into existence, Jean. You can't charm an eel from a pond by watching, just as you couldn't bring your kids home safe by the power of will.

As I passed a house a security light flashed on, and the stretch of lane between me and the lit driveway was solidly dark. Staring into the dark is like encountering your deep self, I thought: the unfathomable source some people call the soul. *Darkness as revelation*, like the cave where the inner eye is opened and the workings of time are fathomed. *Darkness as solace*, with its succour and healing, found in the box of old slides with handwritten notes, or on a piano played by the moon. Then a bend in the road blocked off the bright light, and I saw I was nearly home.

At that moment I noticed a tiny spark on the tarmac a few metres ahead, perhaps imaginary to begin with but growing more real as I approached. Suddenly, just when I'd given up, the night was touched with green magic.

Between

When I was a child, my mum sent me a postcard from hospital, where she was recovering from surgery. It featured a lenticular image – all the rage at the time – of a galleon in heavy seas. The picture looked three-dimensional, and when you tilted it from side to side the waves seemed to move and the sails of the ship to ripple. I was infatuated with it, and desperate to find out how it worked. The card was bonded to a plastic front with the picture printed on it, and it seemed to me that if I could separate the two I would find the magic sandwiched between them. It was one of my most treasured possessions, but eventually curiosity got the better of me and I decided to try. I pushed my two thumbnails between the layers and began to peel them apart.

How can we separate the enigma of real darkness from its negative associations? Those familiar ways of seeing will always be with us; they are as old as human thought itself, and language has grown around them like a tree round barbed wire. We share a primal fear of the dark, because it confronts us with our mortal frailty. Our night vision is limited, and our natural instinct is to find our way to the reassurance of the light. We have a vocabulary, handed down the centuries, which we all draw on to speak of the unspeakable things that frighten and confound us.

But darkness has us under a double spell. It opens us up

not only to fear but also to risk, delight and transformation. Along with the well-worn lexicon of doom, we need ways of expressing its true worth. Yes, it is partial and subjective, an absence that is not an absence, a kind of illusion. But at the same time it could hardly be more real, or more important, since all life depends on it. These are the paradoxical truths so closely bonded that they can't be peeled apart.

Perhaps this is what I've been writing all along: a book-length postcard, to replace the one that ended up in bits all those years ago, when I first tried to separate the inseparable.

Orchard

The wassail is a highlight in the apple-growing year. It takes place in January, when the nights are long and the trees are in their winter dormancy. The wassailers chase off evil spirits, toast the good health of the trees and petition them to fruit copiously the coming year.

This is cider country, and in the past anyone with a bit of ground to spare would have made their own. It's still made that way now, not only as a large-scale commercial business but also by a host of enthusiastic craft producers. Each cider orchard has its wassail, and a couple of years ago we went along and took part in one. It was a spectacular affair, with a huge bonfire, morris dancing, flaming torches and shotguns. The night crackled with good pagan energy, and as we staggered home we said to each other: next year let's have our own.

When we bought the cottage, it came with a small scrubby field which the estate agents called a 'paddock'. The word is generally associated with the keeping of horses, but there was no history of that here and the field would have been quite unsuitable. It was just a scruffy bit of field, steep and rough, with barbed wire fences and the rubble of old chicken coops rotting by the hedge.

There were four trees in the field when we arrived: a cherry, a crab apple, an apple and an unrecognisable other, twisted, stunted and grown through with goat willow that had self-seeded all over the place. The cherry blossomed magnif-icently and was stripped of its cherries by birds and squirrels. The apple produced a few small red fruit, tasty enough, which we ate straight from the tree. We cut away at the goat willow and pulled it up by the roots and freed a second apple, which we nurtured back to productive life.

Now we were hooked. One gauzy summer evening, standing by the gate in the dusk, watching bats work the warm air over the field, we decided we would plant more trees and make it into an orchard. It's easy to dream up these grand schemes in the dusk. Next day we took down the barbed wire and broke up the chicken coops, then sat down with a brochure from the tree nursery and made a plan.

Five years on, most of that original planting has thrived, including some of our more experimental choices – quince, medlar, mulberry. We've had our share of failures, and we've had to accustom ourselves to the sad ritual of digging up a dead fruit tree. We haven't tried making cider yet. But it has started to feel legitimate to call it an orchard.

Last year, when Christmas had come and gone and the January dark was upon us, it was wassail time again. Clearly ours would not be on quite the same scale as the one at the cider place. But in the spirit of living folk tradition we adapted what we'd seen and heard and came up with our own version. We bribed our nearest neighbours with the promise of hot spiced cider, and they agreed to join us, used by now to our eccentricities. Their daughter, crowned with ivy, acted as wassail queen. The dark was like satin, the air still and frosty. The five of us tried to keep warm by processing from tree to tree, which only took about three minutes, and singing in more than one key a traditional wassail song:

Old apple tree, we wassail thee
And hope that thou will bear
Hatfuls, capfuls, three bushel bagfuls,
And a little heap under the stair.

The wassail queen poured a ritual draught of cider around the roots of the largest apple tree in the orchard, and placed a piece of toast in the fork of its branches, as an offering to the visiting robin. We banged pots and pans and hollered at the tops of our voices. We couldn't lay hands on a shotgun, but the neighbours had brought along some dodgy firecrackers, which must surely have scared away any evil spirits lingering in our valley or the next.

Later, the moon rose behind a fuzz of cloud. The beech trees at the top of the orchard cast long shadows on the ground, and we were able to see each other's faces, and to find the mustard for the hot dogs. We stood around together, talking about everything and nothing, feeling the hot cider and the warmth of good company flowing through us in spite of the cold. Every so often one of us would shout *'Was hael!'* and the rest respond *'Drinc hael!'*

People take their wassailing seriously round here, and villages are distinguished by their own particular traditions. But what really makes it is the deep January darkness. This point in the year is ripe for fire and friends and a dose of superstition. And crucially, it worked: come autumn our young trees produced so many hatfuls, capfuls and bagfuls that we had to stack boxes of fruit in the barn, and were still eating them next apple blossom time.

Notes And Acknowledgements

COTTAGE

Part of this chapter first appeared in 'Other Clocks', published in *Hinterland* magazine (issue 11, summer 2022).

Byron wrote 'Darkness' during July and August 1816, sometimes called 'the year without a summer', when the eruption of Mount Tambora sent a huge cloud of volcanic ash into the atmosphere, blocking sunlight and leading to climate disruption. Apocalyptic thinking was fuelled by the prediction of an astronomer in Bologna that the sun would burn itself out on 18 July. But it's hard to separate out the strands of thought and feeling that run through the poem: Byron's personal life was in a state of crisis, and he was staying in a villa on Lake Geneva with the Shelleys where they had challenged one another to write horror stories, the most famous of which is Mary Shelley's novel *Frankenstein.*

Somerset is one in a series of illustrated guide books called Visions of England, published in the late 1940s. In her introduction Sylvia Townsend Warner protests: 'since I am constitutionally incapable of resembling a guide, an err-and-stray-book would be nearer my measure'. Recently released files have revealed that she and her partner Valentine Ackland were under surveillance by MI5 for over twenty years. When I think of Sylvia dashing around the Blackdowns gathering material for her *Somerset* book, I imagine an agent following in a more sober vehicle, trying not to draw attention to his presence by slamming the brakes as he takes the precipitous twists and turns of the lanes.

It's possible that the Blackdown Hills were named for their darkness – the Old English word *'bloec'* means 'bleak' or 'black' – but an alternative source is *'blag don'*, meaning 'wolf hill'.

In his thirties Henry Vaughan underwent a religious conversion, and his poetry took on a mystical dimension. Mystics and visionaries have been drawn to darkness, as a state conducive to spiritual struggle and growth; and light-dark tensions or contradictions are a recurring feature of their writings. Perhaps the greatest examples are in the work of John of the Cross, where darkness is essential and sometimes irresistible This is from his best-known poem, 'The Dark Night' (translated by Martha Sprackland):

> In that night of bliss and joy
> seen by none, in secret and quiet,
> with no clear vision I stepped forth
> with neither guide nor light
> save that which burned within my heart

The Elizabethan playwright and pamphleteer Thomas Nashe wrote *The Terrors of the Night, or, a discourse of apparitions* when he was still in his twenties. It's a satirical attack on demonology and superstition, couched in irrepressible and sometimes lurid style. He is particularly scathing about the interpretation of dreams: 'A dream is nothing else but a bubbling scum or froth of the fancy which the day hath left undigested, or an after-feast made of the fragments of idle imagination'.

Keats was always acutely sensitive to the potential of darkness both actual and figurative. In another letter the following year, he likens life to 'a large mansion' which we feel our way around, and where 'many doors are set open – but all dark – all leading to dark passages'.

SICK ROOM

'Acquainted with the Night' was published in 1928. It's often read as a poem of melancholy or depression, but as in so many of Frost's poems there are folds of ambiguity that are impossible to smooth out: a distant cry stops him in his tracks but seems to have nothing to do with him; the 'luminary clock' of the moon tells a time that is 'neither wrong nor right'; and he is – to the reader as well as the nightwatchman – 'unwilling to explain'.

Erazim Kohák's strange and contemplative book *The Embers and the Stars* contains a chapter called 'The Gift of the Night' which has been a profound influence on my own work. His descriptions of life in a remote part of New Hampshire are so atmospheric that I wanted to go and find the place. Philosophy is braided together with nature writing, and he writes not of conflict between daylight and darkness but of the need to find balance between them. It was in this chapter that I first encountered the idea of pain as a gift: for Kohák it's one of the three gifts of the night, along with darkness and solitude. He is principally interested in psychic rather than physical pain, and the way in which being outdoors alone at night both confronts us with it and changes our perspective on it. 'The grief does not grow less beneath the vast sky,' he writes, 'only it is not reflected back.'

BED

My parents, like so many others of their generation, were followers of Dr Benjamin Spock, whose manual *The Common Sense Book of Baby and Child Care* was first published in 1946. Spock was an advocate of fresh air, and recommended that infants spend a proportion of each day outdoors, summer and winter, warmly wrapped in a pram.

LANDING

Mary Webb's *The House in Dormer Forest* is one of the doomy novels of rural life satirised by Stella Gibbons in *Cold Comfort Farm*. A lot of thought is given these days to the age-appropriateness of books, but there can be a particular pleasure in not understanding; after reading the first page, I puzzled for months over the mysterious phrase 'bats slipped from their purlieus'.

The mother of millions plant has the botanical name *Kalanchoe delagoensis*. Looking at a picture now, and seeing the rows of tiny plantlets along every leaf edge, I feel the same old panic.

The story of the Blackdown Hills Mission is told in *Strength of the Hills* by Ronald H. White. The bookplate in my copy was made out to one Julie Hopkins of St Barnabas Church in 1968. It's a book full of evangelical zeal, and would no doubt have made the perfect Sunday school prize, but I note that the pages fall open not on the stories of salvation and river baptism but on the most lurid scenes of wickedness and ignorance among the heathen locals.

CAVE, 1984

I'm struck by the similarity between my experience of edgelessness and this sentence in Tim Edensor's book *From Light to Dark*: 'Where the gloom thickens, the body's boundaries become indistinct, merging with the surroundings to produce an expansive impression of the space beyond us as we become one with the darkness'.

Reading up on Porth yr Ogof now, I find there are a few intimidating features with names like The Wormhole, The Letterbox and The Washing Machine. I imagine the 'lake' was White Horse Pool, which is said to be several metres deep and to take its name from the shape of an exposed patch of white calcite on the wall above. I don't recognise any of it from the photographs – perhaps that's not surprising given the harrowing nature of the experience, or maybe I'm remembering it wrong and it was somewhere else entirely.

SKY

The Bortle Scale was invented by amateur astronomer John E. Bortle, and was first published in *Sky & Telescope* magazine in 2001.

I am grateful to Jo and Pete Richardson for helping me understand what I see when I look at the night sky. Jo is an astronomer and educator, and Pete is an astrophotographer and telescope maker. Together they run stargazing events on Exmoor, which was Europe's first International Dark Sky Reserve.

Dani Robertson is dark sky officer for the Prosiect Nos

partnership between Snowdonia National Park, the Clwydian Range and Dee Valley, Anglesey and Pen Llŷn areas of outstanding natural beauty.

In his book *The Discarded Image*, C. S. Lewis writes that in the medieval geocentric view of the universe darkness was not infinite but local: 'merely the shadow cast by our Earth'.

CAMPUS

My first Reclaim the Night march would have been in 1980 or 1981, and was part of a movement that began in Leeds in 1977, at the time of the 'Yorkshire Ripper' serial killings and police advice that women should stay indoors at night.

It goes without saying that darkness is not experienced equally. Rebecca Solnit's *Wanderlust* is eloquent on the subject of who is and is not allowed to be out and about at night.

In her seminal book *Pilgrim at Tinker's Creek*, Annie Dillard writes memorably about darkness, at one point noting how it 'pooled in the cleft of the creek and rose, as water collects in a well'. She also remarks: 'If we are blinded by darkness, we are also blinded by light. When too much light falls on everything, a special terror results.'

CAVE, 1962

The conditions in Mammoth Cave were more salubrious than in Scarasson: there's a photograph of a waiter presenting Nathaniel Kleitman and his graduate student Bruce Richardson with a lunch basket and a packet of mail.

My thanks to David Wilson for sharing his research on darkness, trance and mystical experience in Ancient Greece. David also referred me to the work of Yulia Ustinova, whose article 'Cave Experiences and Ancient Greek Oracles' includes this: 'Sensory deprivation is one of the common techniques of inducing altered states of consciousness. They can be attained by different methods, and involve different experiences, but they share a most important characteristic: they silence the waking consciousness and free the mind from the limitations of the alert ego, allowing self-transcendence and awareness undisturbed by the external world.'

A dark retreat can be as short as a weekend, but some people go much further: I have read accounts of six weeks and even three months spent in isolation and total darkness. I was keen to try a dark retreat myself, until I read that after a certain point technicolour imagery takes over and fills every waking moment. That sounds unbearable – I need darkness.

The kinds of learning thought to happen in utero include recognition of the mother, the sucking reflex and language acquisition. There was also a piece of research in the 1990s suggesting that the newborn babies of women who watched *Neighbours* during pregnancy recognised the theme tune.

MUSIC ROOM

Like Valentine Ackland and Sylvia Townsend Warner, the Wordsworths were also under surveillance. They were suspected of fomenting revolution, and walking in the dark was considered a suspicious practice.

In his poem 'Ludwig Rellstab's Visit to Beethoven', Andrew Glaze imagines the scene when Rellstab at last came face-to-face with his hero, only to find him dying, and to have to search for genius in 'that parched face'. Meanwhile, Rellstab's own poems are so bad as to be almost unreadable, though Schubert saw fit to set some of them to music.

T. S. Eliot wrote part three of *The Waste Land* while sitting in the Nayland Rock shelter on the prom at Margate.

Darkness is a recurring presence in Katie Paterson's work. *All the Dead Stars* is a sheet of black steel etched with the locations of 27,000 stars that are no longer visible; and *History of Darkness* is an archive of slides of completely dark areas of the universe in different times and places.

CAVE, 1988

I came across Rebecca Whiteley's *Birth Figures* through a piece called 'Desperate Midwives' by Erin Maglaque. Of an early eighteenth-century engraving depicting a male midwife's hand shining like a light

266

in the darkness of the uterus, she writes: 'These guys really thought that they were enlightening the womb, the last preserve of unreason'.

HOSPITAL

The word 'oxytocin' comes from the Greek for 'swift childbirth'.

LEAD MINE

The Peak District Mining Museum in Matlock Bath is a treasure house of information and artefacts relating to lead mining.

Whether or not mines were ever lit by fireflies, the natural light given off by living things has been pressed into service by humans in various ways: Japanese soldiers in the Second World War read maps by the light of ostracods; bioluminescent bacteria help identify toxins in water supplies; and research is now underway into the use of luminous plants to light buildings and streets.

Nellie Kirkham's book *Derbyshire Lead Mining Through the Centuries* has been an invaluable source of knowledge on the subject, including a detailed account of the Magpie Mine disaster. Nellie was a powerhouse: a highly respected industrial historian, and also a prolific illustrator, author, poet and broadcaster. In one poem, 'The Derbyshire Border', there is a claustrophobic description of the dark sky as a 'cupola clamped down on the disc of the earth'.

Derbyshire lead mines were given wonderful names. Personal favourites include Bacchus Pipe, Beans and Bacon, Cackle Mackle and Wanton Legs.

The ghost of Gamaliel Hall makes an appearance in Roger Flindall's study *Mines, Quarries and Murders in the Peak District*. I wonder whether D. H. Lawrence ever spotted him – Mountain Cottage, where he and Frieda lived in 1918–19, is directly above Goodluck Mine. Lawrence said at the time: 'It is in the darkish Midlands, on the rim of a steep deep valley, looking over the darkish folded hills – exactly the navel of England, and feels exactly that.'

The relationship between the Welsh language and coal mining is a complex and contested one. It's likely that in some areas the influx of English miners contributed to the marginalisation of Welsh.

RIVERBANK

My thanks to Derby Museum and Art Gallery, and to Matt Edwards, the Joseph Wright Study Room curator.

Anna Seward's name appears in Joseph Wright's sitter book, but it is mysteriously struck through. Perhaps she changed her mind; she was ambivalent about the portrait business. 'It was always my resolve never to sit for one between the periods of forty and sixty,' she wrote in a letter to a friend. 'A portrait, where any portion of youthful appearance can be preserved, may be pleasing, and it may be interesting in the mellow tints of venerable age; but the hardness of middle life is detestable on canvas, or ivory.'

Dr David Clarke is co-founder of the Centre for Contemporary Legend at Sheffield Hallam University, and author of *Supernatural Peak District.*

There have always been superstitions associated with moonlight. The novelist Alison Uttley, who was born very close to Dovedale, recalled that when she was a child it was considered unlucky to point at the moon, to look at a new moon through glass or to tread on a moon shadow. On the other hand, the best way to bleach delicate linen was under the light of a full moon.

A 'glamour' was a spell that affected the eyesight, making objects appear different from how they really were. But the word originally derived from 'grammar', and dates back to a time when magic and sorcery were thought to belong with the study of literature.

STREET

Nick Dunn, writing about Manchester at night, describes how 'the shadows refuse to conform to the allocated building plots, skewed, stretched and squeezed across facades and streets alike'.

My thanks to Kerem Asforoglu for his time and patience in explaining about sustainable lighting and the creative initiatives he leads through his company Dark Source. Kerem also introduced me to DarkSky International, and taught me about the five principles for responsible outdoor lighting: it should be useful, targeted, low-level, controlled and warm-coloured.

Craig Koslofsky, author of *Evening's Empire*, writes about lantern smashing as a form of resistance in eighteenth-century cities. The authorities in Vienna resorted to extreme forms of deterrence, threatening to cut off the right hand of any perpetrator they caught.

One of my favourite evocations of the interplay between urban darkness and light comes from Esther Kinsky's novel *River*: 'Now that the rain had turned to the invisible drizzle of a marshland night, the dim light on the glistening asphalt made the surrounding darkness, for all its flickering lights, neon signs and headlamps, seem a kind of ink, a spilt bluish-black, in which countless lives on every side were now immersed.'

I am grateful to Dr Sam Fabian for taking the time to tell me about his research into insects and artificial light.

'Solastalgia' was coined by philosopher Glenn Albrecht in 2003, and 'noctalgia' by astronomers Aparna Venkatesan and John Barentine in 2023.

ISLAND

The man who set off on skis to fetch help for his wife during a difficult labour was the legendary hunter Hilmar Nøis, who built the hut the Ritters lived in and comes to visit them there towards the end of *A Woman in the Polar Night*. Ellen was broken by the experience, left Hilmar and went back to the mainland. Later he married a redoubtable woman called Helfrid, who lived with him on Spitsbergen for many years and wrote a memoir titled *Ishavskvinne* ('Woman of the Polar Ocean').

According to Ole Henrik Magga, the Sami people also have about a thousand different words for reindeer. Meanwhile, researchers at the University of Glasgow found that Scots have 421 terms to describe wintry weather, including 'sneesl' (to begin to rain or snow) and 'skelf' (a large snowflake).

I am grateful to David Barrie for sharing his experiences of sailing at night, and his deep understanding of darkness. His book *Sextant* explores the ancient art of celestial navigation.

POND

Johan Eklöf, a researcher specialising in bats, is frequently reminded that their experience of night is beyond our comprehension. 'The darkness is not the world of humans,' he writes. 'We're only visitors.'

CREVICE

The glow-worm has a cameo role in *Hamlet*, acting as a harbinger of dawn and a sign to the Ghost that it's time to dematerialise: 'The glow-worm shows the matin to be near, / And 'gins to pale his uneffectual fire'. I don't know whether the misgendering is down to the Ghost, or to Shakespeare, or to the scientific understanding at the time.

The entomologist Jean-Henri Fabre made a study of glow-worms, and his observations form a chapter in his book *Wonders of Instinct*, published in 1912. He describes the same pattern of light I saw through the hand lens, but in his own more decorous style: 'The two belts, the exclusive attribute of the marriageable female, are the parts richest in light: to glorify her wedding, the future mother dons her brightest gauds; she lights her two resplendent scarves. But, before that, from the time of the hatching, she had only the modest rush-light of the stern.'

My thanks to Robin Scagell for educating me about glow-worms. His knowledge on the subject is only matched by his curiosity, and he's not averse to a bit of literary detective work. In Dorothy Wordsworth's journal she records three sightings of glow-worms in October 1800, on the 8th, 17th and 20th. October is well outside the glowing season, and having consulted the meteorological records Robin's conclusion is that these were not adult females but larvae, which have their own fainter and more intermittent luminescence.

At the height of my glow-worm infatuation I was seeing them everywhere: on a toothbrush, an alarm clock, a television on standby. Our electric car has a charging light that blinks every ten seconds, and I'm still considering sticking a plaster over it so that passing males will not be led astray.

ORCHARD

Like other ancient folk traditions, wassailing was taken up with enthusiasm during the Victorian era, and I make no claim for the authenticity of the wassail song, the toast or my recipe for hot spiced cider.

I am grateful to all those who provided help, advice and encouragement during the writing of this book, including the Baker family, Mandy Coe, Paul Cutts, Emma Harding, Caroline Hawkridge, Kari Leibowitz, Rachel Lichtenstein, Michael Symmons Roberts and Joar Vittersø; and to Bea Hemming and everyone at Cape.

Special thanks to Nigel Pantling for all the support, enthusiasm, wisdom and walks in the dark.